PLATELET-ACTIVATING FACTOR ANTAGONISTS:

NEW DEVELOPMENTS FOR CLINICAL APPLICATION

ADVANCES IN APPLIED BIOTECHNOLOGY SERIES
VOLUME 9

PAF Antagonists: New Developments for Clinical Application

Library of Congress Cataloging-in-Publication Data
PAF antagonists: new developments for clinical application
editors, Joseph T. O'Flaherty and Peter Ramwell
p. cm.—(advances in applied biotechnology series; V. 9)
Includes bibliographical references
Includes index
ISBN 0-943255-13-9
1. Platelet activating factor. 2. Platelet activating factor–Antagonists.
I. O'Flaherty, Joseph T., 1943- . II. Ramwell, Peter W., 1930- . III. Series.
[DNLM: 1. Platelet Activating Factor–analysis. 2. Platelet Activating Factor–pharmacology. 3. Platelet Activating Factor–physiology.
4. Platelet Activating Factor–receptor.]
QP752.P62P34 1990
616.07'95—dc20
DLC
for Library of Congress

Series ISBN 0-943255-08-2

ISBN 0-943255-13-9

On the Cover

Animal lung section showing an infiltration of eosinophils following antigen challenge (left), and the lung morphology observed in antigen-challenged animals after treatment with the PAF antagonist, BN52730 (right).
Color photographs
Photos provided by: Dr. Jean Michel Mencia-Huerta of the Institut Henri Beaufour in Les Ulis, France.

Table of Contents

Analysis of Platelet-Activating Factor

PAF – Receptor Interactions

Foreword

This Conference was convened by International Business Communications (IBC) to discuss the pathophysiology of the platelet-activating factor (PAF), and specifically to explore the clinical indications for PAF antagonists; consequently, most of the individuals attending the Conference were from the pharmaceutical industry.

It should be noted that PAF is a lipid mediator and this is important in many ways for it is very similar to thromboxane A_2 and the leukotrienes. They exhibit a wide range of biological activity as well as being ubiquitous in their distribution which always makes recognition of specific clinical indications difficult. The key to identifying their putative pathophysiologic roles of course has been the highly successful development of potent and specific receptor antagonists. The relationship between PAF and these two eicosanoids is a constant theme since many of PAF's effects can be blocked by thromboxane and leukotriene synthesis inhibitors and receptor antagonists. At this time, however, less progress has been made in developing inhibitors of PAF synthesis.

As with leukotriene antagonists it is not easy to identify convincing indications for PAF antagonists. Asthma is clearly the primary indication, but several other indications were considered at this Conference including septic shock, transplantation, CNS injury, some type of inflammation and cyclosporine-induced nephrotoxicity. As might be expected Phase II-III IND studies are being undertaken with respect to small airway disease.

A large number and variety of compounds have been successfully synthesized to obtain specific PAF antagonists which are orally active. However, some compounds in Phase I are not highly specific PAF antagonists since it is thought that a broader and less specific spectrum of antagonist activity, for example against histamine or leukotrienes as well as PAF, may be advantageous in diseases involving a variety of pathophysiological mediators.

Now, the physiological roles of PAF are arousing interest especially in nidation, lung maturation, and promotion of the immune response. However, the federal agencies and Industry need to encourage further exploratory research since much more information is still needed to provide a basis for more meaningful clinical studies.

Peter W. Ramwell, Ph.D.
Department of Physiology and Biophysics
Georgetown University
3900 Reservoir Road NW.
Washington, DC 20007

Platelet-activating factor (PAF) has far-reaching biological properties. Consider the following points. First, PAF activates thrombocytes, leukocytes, vascular endothelium, macrophages, smooth muscle, neurocytes, glandular cells, dermatocytes, and other tissues. Few, if any, organs are indifferent to the agonist's stimulating actions. Second, these same cell types, when stimulated with any one of various agents, produce PAF. The molecule may form at virtually any site of perturbation. Third, PAF stimulates target cells to release diverse bioactive principles including serotonin, histamine, tumor

Acknowledgments

PAF Antagonists: New Developments for Clinical Application is the compilation of papers presented at the International Conference on PAF and PAF Antagonists, held December 1989 in New Orleans, LA. This Conference is one of many Conferences on Biotechnology held each year and sponsored by International Business Communications, IBC USA Conferences Inc., 8 Pleasant Street, Building D, South Natick, MA 01760.

Two additional papers, not presented at this Conference, were submitted for this Volume by Dr. William L. Salzer of Wake Forest University and Dr. Giora Feuerstein and colleagues at the Department of Pharmacology, SmithKline Beecham Pharmaceuticals. We greatly appreciate their added contributions.

Portfolio Publishing Company wishes to express its gratitude to the staff of IBC for their continued support and cooperation in allowing these excellent compilations to be included in our Series on Advances in Applied Biotechnology. Papers delivered at other IBC Conferences were included in Volume 2, *Discoveries in Antisense Nucleic Acids*, of this Series and will be included in forthcoming volumes on Protein C and Related Anticoagulants and on Technologies and Strategies for Fat and Cholesterol Reduction in Food.

Analysis of Platelet-Activating Factor

Introduction

The platelet-activating factor (PAF) is a potent lipid autocoid first described nearly 30 years ago[1] that is involved in the etiology of a number of diseases as well as basic physiological responses. In fact, the role for PAF is immense, encompassing the entire life cycle. PAF is not only required for critical aspects of conception,[2] it also mediates a major cause of death, septic shock.[3] However, assigning a causative role to PAF has required quantitation, which has been problematic. Until recently, quantitation has relied on insensitive physicochemical or sensitive but nonspecific biological methods of PAF analysis. Current methodology will be divided into five areas and discussed with respect to their strengths and weaknesses.

Bioassays and Enzymatic Assays of PAF

The original and most popular method of PAF analysis is platelet activation.[4] The relative sensitivity of this and the other assays discussed are shown in Figure 1. Typically, the release of serotonin or aggregation yield a log-linear response to PAF sensitive to the picogram level. Although the assay has been critical to the understanding of PAF, results have been confounded by a number of factors, including the presence of natural PAF antagonists, modulators of signal transduction, and lytic factors. Sphingomyelin[5] is a particularly menacing source of interference, because it is abundant and not easily separated by thin-layer chromatography or normal phase HPLC.

A useful complement to the bioassays for PAF is acetate incorporation,[6,7] which relies on the PAF biosynthetic apparatus to synthesize AcetylCoA and transfer (lysoPAF:acetylCoA transferase) the acetate to nascent lysoPAF. PAF is then isolated and assayed for tritium. Because acetate in high-specific activity is available, the synthesis of pg amounts of PAF can be detected. This assay is inexpensive and has been successfully employed by a number of

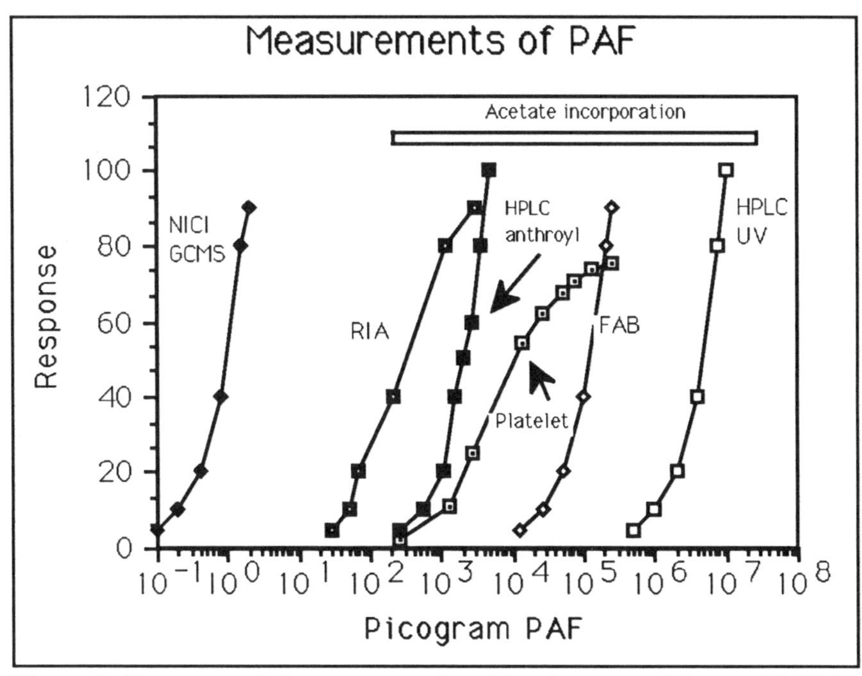

Figure 1. Shown are relative responses (y-axis) to incremental doses of PAF (x-axis). Because acetate incorporation measures rates, not absolute amounts, the range of sensitivity is indicated with a bar across the top.

laboratories for many years. However, there are a number of limitations. Active biosynthesis is required, and therefore; acetate incorporation yields information about rates rather than absolute amounts. Furthermore, biosynthesis must occur by the reacylation-deacylation pathway rather than by *de novo* synthesis.

Quantitation of PAF by Mass Spectrometry

Because of these problems, a sensitive physicochemical method of PAF analysis was required to more clearly define the physiological and pathophysiological roles of PAF. However, inherent characteristics of PAF hinder such analysis at low levels. First, the phospholipid nature of PAF presents poor chromatographic characteristics requir-

homeostatic processes regulating PAF concentrations by comparing the distribution of molecular species in PAF precursors and metabolites.[13-15]

The same analytical strategy was applied to the immediate precursor of PAF, 1-O-alkyl-2-acyl-*sn*-glycero-3-phosphocholine (Alkyl-PC), and the immediate metabolite, lysoPAF. In this case, both of these molecules were mildly hydrolyzed to remove the contaminating 1-acyl-2-acetyl-*sn*-glycero-3-phosphocholine (1-acyl-PAF) prior to enzymatic hydrolysis to the diglyceride. The samples were then acetylated to yield PAF and ultimately converted to the PFB ester for analysis.

It should be noted that useful advancements of this original method have appeared in the literature. First, PAF labeled with deuterium in the alkyl chain has recently become commercially available. This is quite advantageous over $^2H_3$[acetyl]-PAF because of stability to the mild hydrolytic steps designed to remove "1-acyl-PAF" contaminants which are isobaric to PAF. Second, a single-step procedure for the hydrolysis and derivatization of PAF has been described.[16] Finally, analysis of the 1,3-PFB-DG[17] and an improved PAF purification method compatible with this procedure have been described and shown to be particularly useful for the analysis of PAF in human plasma.[18]

HPLC Analysis of PAF

Alternatively, the 1-O-alkyl-2-acetylglycerol can be acylated with anthranitrile to the anthroyl ester shown in Figure 4 for HPLC analysis. This anthra adduct has an extremely high molar extinction coefficient (>60,000), thus permitting detection as low as 50pg. Under reverse phase conditions, molecular species of PAF, lysoPAF, and precursor phospholipids as well as natural diglyceride intermediates can be resolved. Because HPLC instrumentation is less expensive than the mass spectrometric technique, and has no limitation with respect to volatility, this method is quite promising when greater

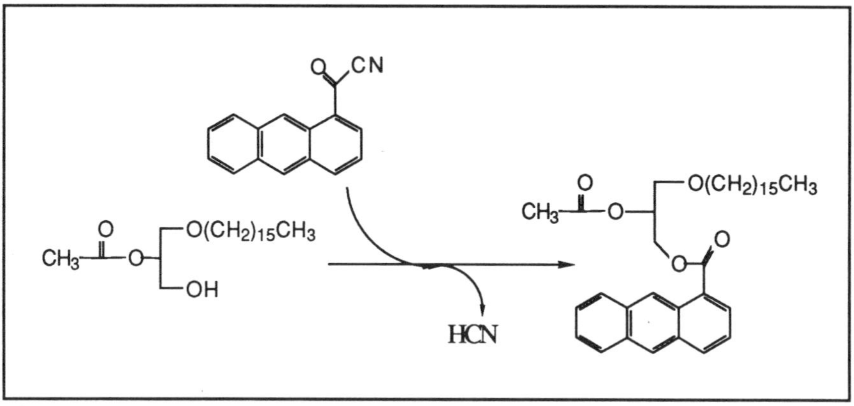

Figure 4.

sensitivity is not required. Because the anthra adduct is highly fluorescent, it is expected that the sensitivity will improve when fluorescent detection systems are explored.

Radioimmunoassay of PAF

The most recent advance in PAF analysis has been the preparation of antibodies. The preparation of antibodies to PAF has been elusive due to the inherent lack of antigenicity and functional groups needed for conjugation. However, the synthesis of an aldehyde at either carbon 6 or 12 of the 1-O-alkyl chain yields a successful immunogen when appropriately coupled.[19]

The commercial radioimmunoassay (RIA) from Dupont has been examined. Linear displacement curves were generated to 50pg. For whole blood, samples from the RIA correlated well with mass spectrometric assays, yielding comparable but consistently greater values. The most troublesome sources of interference are sphingomyelin and "1-acyl-PAF." Because exogenous sphingomyelin causes a similar shift in the curve, it is preliminarily assumed to be the interfering substance. In conclusion, the RIA recently available could be a valuable tool for the analysis of PAF, but it should not be

2

Receptor Heterogeneity and the Existence of Intracellular Receptors of Platelet-Activating Factor

correspondence
San-Bao Hwang
Su Wang
Merck Sharp & Dohme
Research Laboratories
Department of
Biochemical Regulation
P.O. Box 2000 (80B7)
Rahway, NJ 07065

Platelet-activating factor (PAF) is a phospholipid that has diverse and potent effects on many cells and tissues. A cell surface receptor that specifically binds PAF and mediates various responses to PAF both *in vitro* and *in vivo* has been identified in several types of cells and tissues. PAF receptors and G-proteins coupled to PAF receptors in human platelets are different from those in human polymorphonuclear leukocytes. Receptor heterogeneity thus does exist in the PAF receptor system. Also, there exists intracellular PAF receptors that may mediate the responses of PAF produced and retained inside the stimulated cells.

Introduction

Platelet-activating factor (PAF), identified as 1-0-alkyl-2-0-acetyl-*sn*-glycero-3-phosphorylcholine (Figure 1),[1-3] is a phospholipid that has diverse and potent effects on many cells and tissues (reviewed in References 4 and 5). PAF was initially identified as a platelet aggregating soluble product released by immunoglobulin E

maximal detectable receptor number of from 50 (S.-B. Hwang and S. Wang, unpublished results, 1990) to 1400 bindings sites[12] per human platelets.

The specific binding of [^{3}H]PAF was reversible, because [^{3}H]PAF bound specifically to cells[18,23,25] or isolated membranes[17] could be dissociated effectively by an excess of unlabeled PAF. The PAF receptor was functionally active. Relative potencies of PAF and several structure-related analogues both *in vitro* and *in vivo*[31,32] correlate very well with their relative inhibition of [^{3}H]PAF binding to specific receptor sites. Rat platelets do not aggregate in the presence of PAF, nor were red blood cells found to be responsive to PAF; neither of these have a high affinity for PAF receptors.[14,17,33] More important, several PAF receptor antagonists have been identified in the past few years (reviewed in References 34 to 36). These antagonists include natural products and synthetic compounds. These receptor antagonists are all reported to compete with the binding of [^{3}H]PAF to rabbit,[31,33,37] guinea pig,[38] and human platelets[33,38] as well as the binding to human polymorphonuclear leukocytes (PMN),[16] human lung tissues,[26,33] and rat liver tissues.[27] Excellent structure-activity relationship of these receptor antagonists in inhibiting either [^{3}H]PAF binding or induced function of PAF has been well established.[36,39,40] Furthermore, with the tritium labeled PAF receptor antagonists, these compounds were demonstrated to share a common binding site with PAF.[31,37,41-44] These results strongly suggest that binding sites for [^{3}H]PAF truely represent the pharmacologically relevant receptor sites.

Heterogeneity of PAF Receptors

Using isolated membranes, we have observed a single type of PAF-specific receptor in rabbit[17] and human platelets,[33] human PMNs,[16] human lung tissue,[26] rat liver tissue,[27] and rat peritoneal PMNs, with an equilibrium dissociation constant of approximately 0.5nM under identical ionic conditions (Table 1). In a comparison of the oral potency of several leading PAF antagonists, including SRI 63-441,

Table 1
Equilibrium Dissociation Constants (K_D) and Maximal Number of PAF Receptor Sites Determined with [^{3}H]C_{16}-PAF in Isolated Membranes

Membranes	K_d(nM) (± S.D.)	B_{max} (± S.D.) (10^{-13} mol/mg protein)	References
Rabbit platelets	0.53 (0.06)	23.0 (4.0)	45
Human platelets	0.40 (0.10)	2.72 (1.20)	33
Human PMNs	0.47 (0.14)	3.13 (1.40)	16
Human lung tissues	0.51 (0.17)	1.40 (0.37)	26
Rat liver tissues	0.51 (0.14)	1.41 (0.18)	27
Rat peritoneal PMNs	0.61 (0.10)	3.32 (1.34) x 10^4 recepters/cell	
Rat platelets	Not detected		33

BN 52021, 48740 RP, L-652,731, SRI 63-072, and CV-3988, the order of potency for four orally active antagonists of these six , L-652,731 > BN 52021 > 48740 RP > SRI 63-072, observed for inhibition of PAF-induced hypotension in the rat is similar to that for inhibition of PAF-induced hemoconcentration and bronchoconstriction in the guinea pig.[35] Therefore, these antagonists seem to have the same PAF receptor in common.

In contrast, in studies of the specific binding of a tritium-labeled nonmetabolizable PAF analogue, 1-0-hexadecyl-(1',2'-^{3}H)-2-N-methylcarbamyl-*sn*-glyceryl-3-phosphorylcholine (N-methylcarbamyl PAF), the K_d value is significantly different for human platelet and human PML membranes. The mean ± S.D. K_d value for [^{3}H]N-methylcarbamyl PAF in human platelet membranes is 2.18±0.48nM (n=8), which is significantly different (P<0.0001) from the K_d value in human PML membranes (5.18 ± 0.54nM, n=8).[45]

We also compared the relative potencies of several selected PAF agonists and PAF receptor antagonists for human platelet membranes and human PMN membranes; two receptor antagonists (Ono-6240 and FR 72112) were found to have significant differences in potency in binding assay as well as in functional studies.[16,45] Ono-6240 is approximately six times more potent in human platelet (equilibrium inhibition constant (K_i = 4.6 x 10^{-8}M) than in human PMN membranes

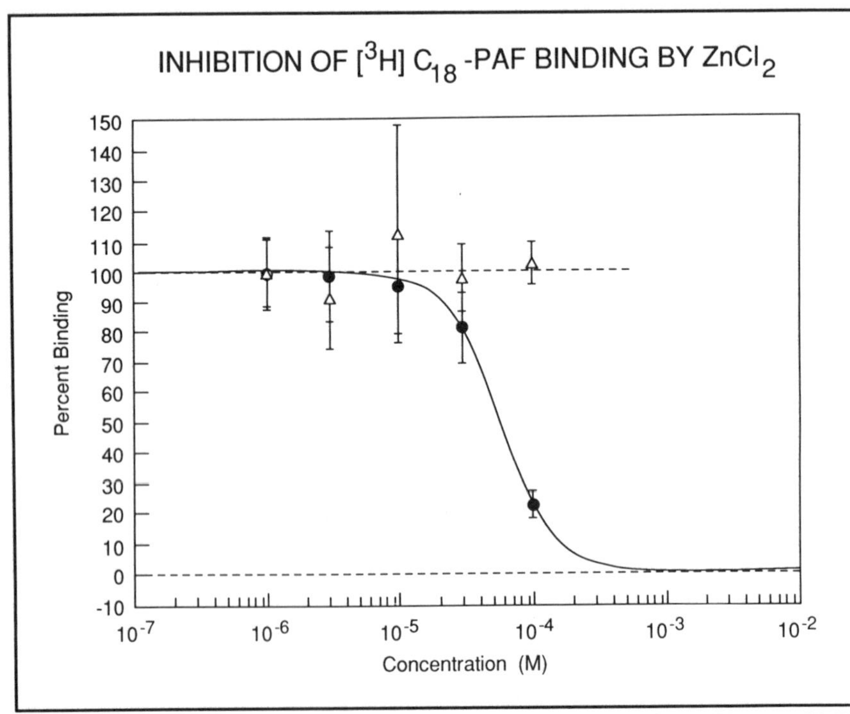

Figure 5. Effect of $ZnCl_2$ on [^{3}H]PAF binding to human platelet and human PMN membranes. Each percentage of inhibition is average of results from three to four independent experiments, and in each experiment, each sample is triplicated. (●—●) = human platelet membranes; (Δ—Δ) = human PMN membranes.

rabbit platelets, human platelets, and human PMNs by Ono-6240. Although the inhibitory potency of Ono-6240 toward PAF-induced aggregation of gel-filtered human platelets was identical to that of rabbit platelets ($K_b = 3.4 \times 10^{-7}$M), Ono-6240 is significantly less potent in inhibiting the PAF-induced human PMN aggregation. The K_b value of Ono-6240 for inhibition of PAF-induced PMN aggregation was 2×10^{-6}M, which is approximately sixfold higher than the corresponding value for inhibition of platelet aggregation.[16]

Na^+ specifically inhibits the binding of [^{3}H]PAF to either rabbit or human platelet membranes, with an ED_{50} value of 6mM.[47] Binding is also inhibited by Li^+ but at a much higher concentration (ED_{50} = 150mM).[47] In contrast, K^+, Cs^+, and Rb^+ as well as Mg^{2+}, Ca^{2+}, and Mn^{2+} enhance [^{3}H]PAF binding.[47] From both Scatchard and Klotz analyses, the inhibitory effect of Na^+ is apparently due to an increase in equilibrium dissociation constant of PAF binding to its receptors. In contrast, the Mg^{2+}-induced enhancement of specific PAF binding arises from both an increased affinity of the receptor and an increased number of receptor sites.[47] The variation in the detectable receptor number under different ionic conditions has been attributed to the coexistence of several conformational states of PAF receptors as concluded from the binding studies of a tritium-labeled PAF receptor antagonist, L-659,989 ([^{3}H]L-659,989).[44] However, no significant inhibition by Na^+ or Li^+ on the specific [^{3}H]PAF binding in human PMN membranes is observed, even at a concentration as high as 300mM.[16] Furthermore, although in both rabbit[47] and human platelet membrane,[16] K^+ exhibits no effect on Mg^{2+}-potentiated specific [^{3}H] binding and K^+ decreases the Mg^{2+}-potentiated [^{3}H]PAF binding in human PMN membranes.[16] Receptor heterogeneity is therefore clearly demonstrated by the differences in (1) the K_d values of [^{3}H]N-methylcarbamyl PAF, (2) the K_i values of the PAF receptor antagonists Ono-6240 and FR 72112, and (3) the ionic modulation of the specific [^{3}H]PAF binding observed between PAF receptors in human platelets and PMNs membranes.

Coupling of PAF Receptors to G-Protein

Guanosine 5'-triphosphate (GTP) specifically inhibits PAF binding to rabbit platelet[47] or human PML membranes.[16] Adenosine 5'-triphosphate and guanosine diphosphate, at a similar concentration, exhibit no inhibitory effect on [^{3}H]PAF binding.[47] These results suggest that the signal transduction process induced by PAF may be mediated by G-proteins. Further evidence to support the coupling of PAF receptors to G-proteins arises from the measurements of PAF-induced GTPase activity in rabbit and human platelets and human PMNs. PAF-stimulated GTPase activity in human[16] and rabbit platelet[47] and in human PMN membranes[16] with an ED_{50} value near 1nM, whereas the biologically inactive enantiomer of PAF (enantio-C_{16}-PAF) does not stimulate GTP hydrolysis even at micromolar concentrations.[16,47] Such a PAF-induced GTPase activity can be specifically inhibited by the PAF-specific receptor antagonists kadsurenone[47] or Ono-6240[16] but not by the inactive analogue.[47] These results strongly support the hypothesis that PAF-induced GTPase activity is a receptor-mediated process.

Bacterial toxins have proved to be useful tools to distinguish among G-protein types. The α-subunit that binds guanine nucleotides with affinities in the submicromolar range is a GTPase and serves as a substrate for adenosine 5'-diphosphate ribosylation by bacterial toxins. The α-subunit of each G-protein is distinct and may determine the specificity of a given G-protein for receptor and effector coupling. With the help of pertussis toxin and cholera toxin, we have demonstrated that the G-protein coupled to the PAF receptor in human platelets is different from that in human PMNs.[16] Pertussis toxin and cholera toxin totally abolish PAF-stimulated GTPase activity in human neutrophils, whereas neither toxin has an effect on human platelets. Similarly, treatment of human platelet membranes with either pertussis toxin or cholera toxin has been reported to have little or no effect on the GTPase activity stimulated by PAF,[48] and pertussis toxin inhibits PAF-mediated chemotaxis, superoxide production, aggregation, and release of lysozyme contents in human neutrophils.[49] Therefore, both the PAF receptors and the G-proteins coupled to the receptors are structurally different in human platelets and neutrophils.

Existence of the Intracellular PAF Receptor

Synthesis of PAF occurs concomitantly with cellular activation. Considerable quantities of intracellularly produced PAF remain within many cells.[8,11] Such retention led to the speculation that PAF could act as both an intracellular and extracellular mediator. An interesting recent model was proposed that the intracellularly retained PAF may act as fusogen,[50] altering the physical properties of membranes and thereby affecting membrane-bound proteins or ionic channels. However, although PAF has been shown to fluidize the membrane hydrophobic core and lower both the pretransition and phase transition in vesicles of dipalmitoyl phosphorylcholine,[17,51] a high mole ratio of PAF to synthetic phospholipid was required to effect a significant change in the transition temperature or to broaden the transition width. In addition, no significant difference was observed in the effect of PAF and the 1-acyl analogue of PAF (1-palmitoyl-2-acetyl-*sn*-glyceryl-3-phosphorylcholine) on several physical properties of pure dipalmitoyl phosphorylcholine bilayers as determined by differential scanning calorimetry,[17] even though PAF is approximately 1000 times more potent than the 1-acyl PAF analogue in inhibiting the specific [^{3}H]PAF binding to rabbit platelet membranes and in inducing aggregation of gel-filtered rabbit platelets.[17,31] These observations argue against the proposed action of PAF as an intracellular fusogen.

Specific binding of [^{3}H]PAF or [^{3}H]N-methylcarbamyl PAF to human platelet membranes was found to be potentiated specifically by wheat germ agglutinin (WGA) or erythroagglutinin (PHA-E), whereas concanavalin A from *Canavalia ensiformis* (Con A), lectin from *Arachis hypogaea* (PNA), lectin from *Ulex europaeus* (Gorse), and lectin from *Bandeiraea simplicifolia* (BS-I) at similar dose ratios showed no effects on the binding (Figure 6, A). The potentiation effect as demonstrated in Scatchard plots was due to an increase in the detectable number of receptor sites but not to an alteration of the equilibrium dissociation constant of the radioligand from receptors (Figure 7 and Table 2). N-acetyl-glucosamine specifically inhibits the enhancement with an ED_{50} of 7mM (Figure 8). D-glucosamine and N-acetyl-galactosamine are less potent, whereas L-fucose at similar concentrations does not inhibit the potentiated binding to human platelet membranes.

C_{16}-PAF and L-659,989 fully displace the binding of [^{3}H]N-methylcarbamyl PAF to human platelet membranes with identical equilibrium inhibition constant either in the presence or absence of 30μg WGA/ml (Table 3). The WGA-potentiated binding was also inhibited by Na^+ in an identical manner to the binding of [^{3}H] N-methylcarbamyl PAF in the absence of WGA (Figure 9). These results suggest that the PAF receptors exposed by WGA are identical to the PAF receptors present in the absence of WGA. Binding of WGA to a

Figure 6. Potentiation effects of [^{3}H]N-methylcarbamylPAF binding to membrane fraction A (**A**) and membrane fraction B (**B**). Lectins were added to an incubation mixture containing 100mg membrane protein, 1nM [^{3}H]N-methylcarbamylPAF, in 10mM $MgCl_2$, 10mM Tris, and 0.25 percent BSA at pH 7.0. Data were normalized so that specific binding in absence of lectin is 100 percent. Error bar is standard deviation of four independent experiments. In each experiment, quadruplicate determinations were performed.

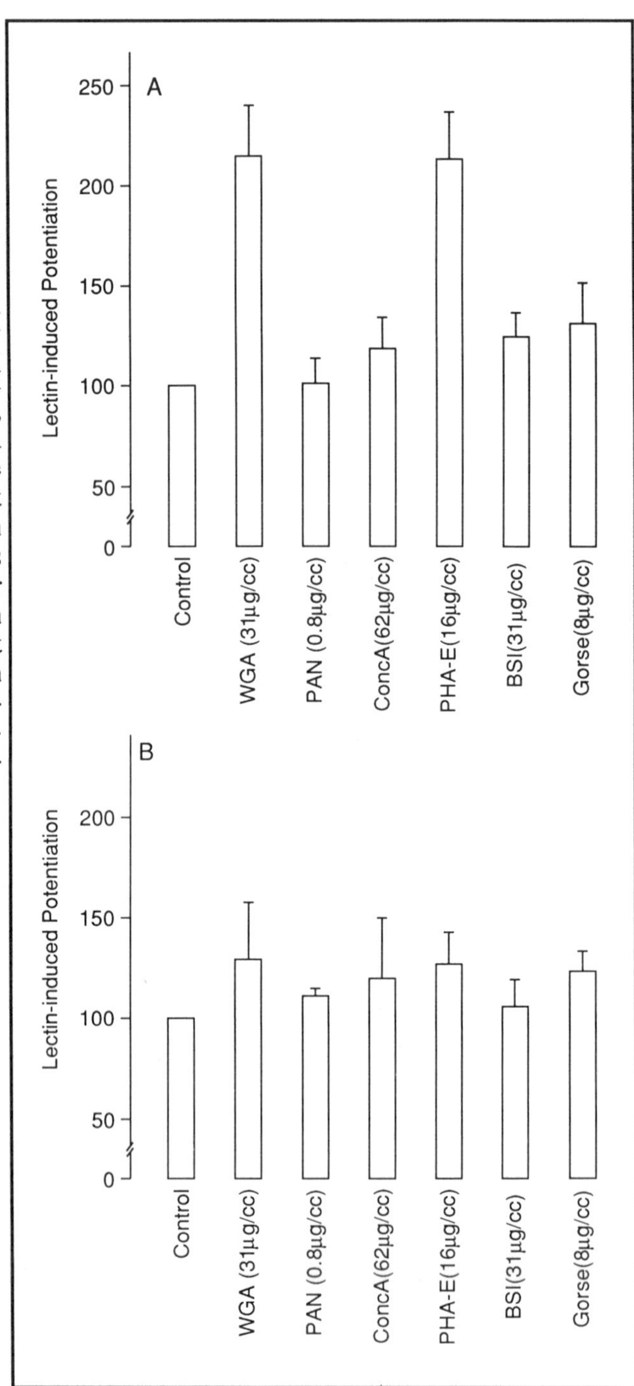

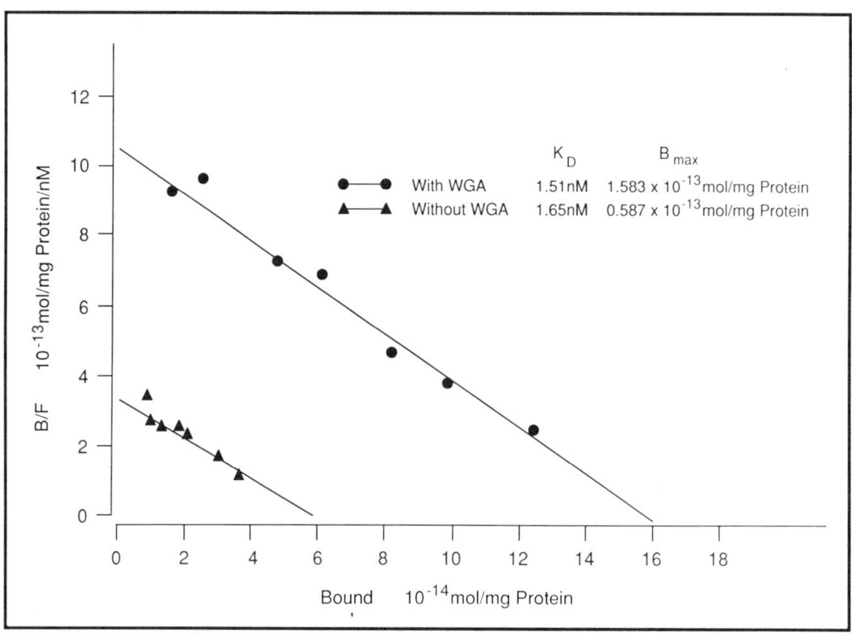

Figure 7. Scatchard plots of specific [^{3}H]N-methylcarbamyl PAF binding in presence (●—●) and in the absence (▲—▲) of 30μg WGA/ml. 100μg membrane protein were added to tubes containing [^{3}H]N-methylcarbamyl PAF ranging from 0.2 to 6nM in an assay medium of 10mM $MgCl_2$, 10mM Tris, and 0.25 percent BSA at pH 7.0. Each data point is mean of quadruplicate determinations.

Table 2

Equilibrium Dissociation Constants (K_d) and Maximal Binding Sites (B_{max}) of Tritium-Labeled C_{18}-PAF and N-Methylcarbamyl PAF in Human Platelet Membranes

Ligand	Membrane Number	WGA* (μg/ml)	K_d (± S.D.) (nM)	B_{max} (10^{-14}mol/mg protein)
N-methyl-carbamyl PAF	334A	30	1.98 (0.17[1])	9.23 ± 0.22 (*n*=2)
		0	1.69 (0.26[1])	3.11 ± 0.59 (*n*=4)
	280A	30	1.55 (0.06[2])	18.34 ± 2.4 (*n*=4)
		0	1.53 (0.13[2])	5.71 ± 0.3 (*n*=4)
C_{18}-PAF	334A	30	1.62 (0.27[3])	10.55 ± 0.95 (*n*=3)
		0	1.54 (0.04[3])	5.45 ± 1.1 (*n*=3)
	306A	30	1.75 (0.28[4])	30.6 ± 7.4 (*n*=4)
		0	1.72 (0.24[4])	14.1 ± 2.5 (*n*=4)

[1]*P* = 0.23, not significant.
[2]*P* = 0.81, not significant.
[3]*P* = 0.63, not significant.
[4]*P* = 0.80, not significant.
*WGA = wheat germ agglutinin.

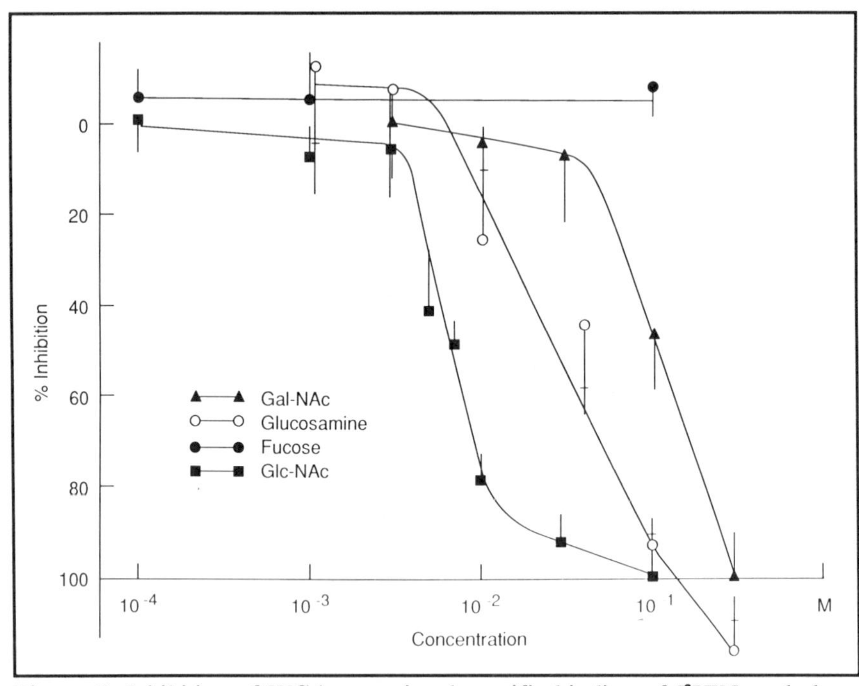

Figure 8. Inhibition of WGA-potentiated specific binding of [^{3}H]N-methylcarbamyl PAF by N-acetyl-D-glucosamine (■—■), D-glucosamine (○—○), L-fucose (●—●), and N-acetyl-D-galactosamine (▲—▲). Assay conditions were identical to those described in Figure 6. Data were normalized so that specific binding with 30μg WGA/ml was defined as 0 percent inhibition, and specific binding in absence of WGA is 100 percent inhibition. Data point and error bar are mean and standard deviation of three to four repeated experiments, respectively.

Table 3

Equilibrium Inhibition Constants (K_I) of C_{16}-PAF, Enantio-C_{16}-PAF, and L-659, 989 in Inhibiting the Binding of [^{3}H]N-Methylcarbamyl PAF to Human Platelet Membranes

	K_I(M)	
Compound	**Without WGA* (± S.D.)**	**With 30μg WGA/ml**
C_{16}-PAF	7.33 (2.28) x 10^{-10}	9.44 (0.97) x 10^{-10} (*n*=3)
Enantio-C_{16}-PAF	2.32 (0.57) x 10^{-6}	3.13 (1.0) x 10^{-6} (*n*=3)
L-659, 989	6.59 (4.9) x 10^{-9}	4.72 (1.6) x 10^{-9} (*n*=3)

*WGA = wheat germ agglutinin.

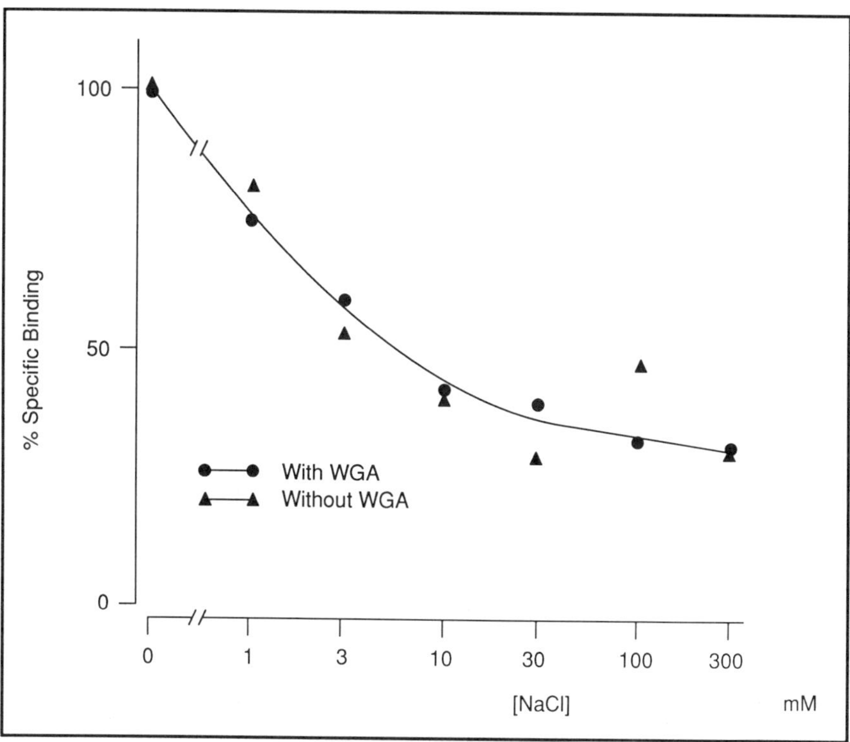

Figure 9. Inhibition of specific [^{3}H]N-methylcarbamyl PAF binding to human platelet membranes by Na^+ in 10mM $MgCl_2$, 10mM Tris, and 0.25 percent BSA at pH 7.0 with (●—●) or without (▲—▲) 30μg WGA/ml. Data were normalized so that specific binding of [^{3}H]N-methylcarbamyl PAF in absence of Na^+ is 100 percent.

membrane component possessing terminal N-acetylglucosamine residues transformed those PAF receptors in low affinity state(s), which is not detectable by the conventional radioligand binding assay, into high affinity state(s).

Human platelet membranes prepared by the freezing and thawing procedure of Hwang *et al.*[17] were routinely separable into two fractions on a discontinuous sucrose density gradient (Figure 10). The major portion of membrane preparations (membrane fraction B) was found at the interface between 1.03 and 1.5M sucrose and contained more receptor sites for PAF than membrane fraction A, which was found at the interface between 0.25 and 1.03M sucrose.[32] No detect-

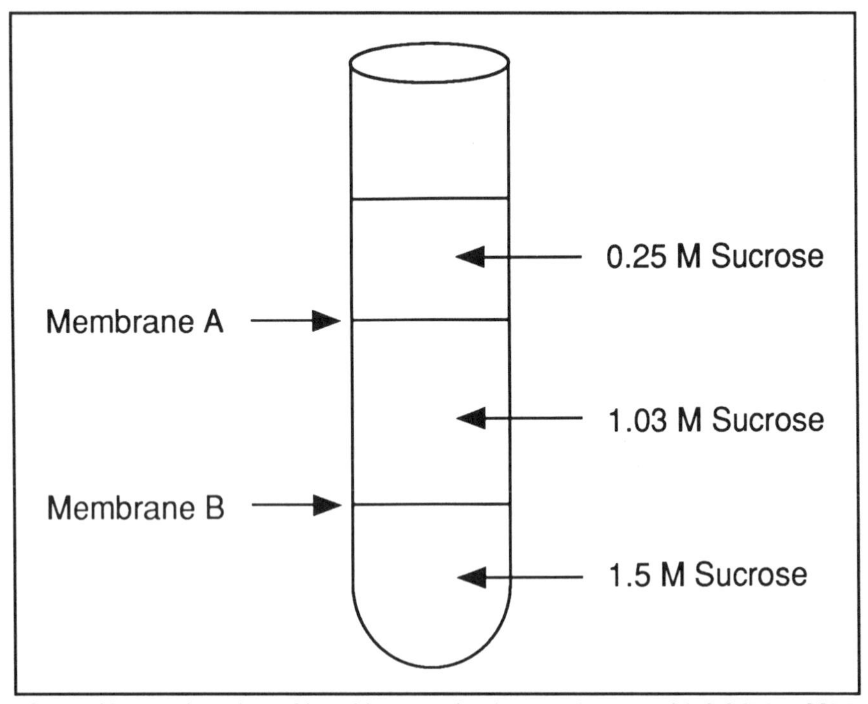

Figure 10. Fractionation of lysed human platelet membranes with 0.25M, 1.03M, and 1.5M sucrose discontinuous density gradient.

able differences were found in the K_d values of [^{3}H]PAF, [^{3}H]N-methylcarbamyl PAF, and [^{3}H]L-659,989 or in the K_i values of several receptor antagonists in the PAF receptor of membrane fraction A and membrane fraction B.[32] However, the potentiation effects induced by WGA or PHA-E were found to be more pronounced in membrane fraction A than in membrane fraction B (Figure 6, B). Also, membrane fraction A exhibited greater alkaline phosphatase activity than did membrane fraction B,[41] whereas membrane fraction B contained more receptors for inositol-(1,4,5)-trisphosphate ($InsP_3$) than did membrane fraction A (S.-B. Hwang, unpublished results, 1990) and membrane fraction B was selectively enriched with antimycin-insensitive nicotinamide adenine dinucleotide-cytochrome C reductase. Therefore, membrane fraction A is a likely plasma membrane-enriched fraction, whereas membrane fraction B is enriched with intracellular membranes. Because the intracellular membrane fraction (membrane fraction B) consistently contains more PAF receptor sites than the plasma membrane-enriched fraction,[33] it is clear that intracellular membranes may contain PAF receptors. These intracellular PAF receptors may initiate cellular responses to PAF synthesized and retained intracellularly.

References

1. C.A. Demopoulos, R.N. Pinckard, and D.J. Hanahan, *J. Biol. Chem.* **254**, 9355 (1979).
2. J. Benveniste *et al., C.R. Hebd. Seances Acad. Sci. Ser. D Sci. Nat.* **289**, 1037 (1979).
3. M.L. Blank *et al., Biochem. Biophys. Res. Commun.* **90**, 1194 (1979).
4. P. Braquet et al., *Pharmacol. Rev.* **39**, 97 (1987).
5. F. Snyder, *Proc. Soc. Exp. Biol. Med.* **190**, 125 (1989).
6. J. Benveniste, P.M. Henson, and C.G. Cochrane, *J. Exp. Med.* **136**, 1356 (1972).
7. B. Armoux, D. Duval, and J. Benveniste, *Eur. J. Clin. Invest.* **10**, 437 (1980).
8. S.J. Bentz and P.M. Henson, *J. Immunol.* **125**, 2756 (1980).
9. M. Chignard *et al., Br. J. Haematol.* **46**, 455 (1980).
10. T.-C. Lee *et al., J. Biol. Chem.* **259**, 5526 (1984).
11. S.M. Prescott, G.A. Zimmerman, and T.M. McIntyre, *Proc. Natl. Acad. Sci. USA* **81**, 3534 (1984).
12. F.H. Valone *et al., J. Immunol.* **129**, 1637 (1982).
13. E. Kloprogge and J.W.N. Akkerman, *Biochem. J.* **223**, 901 (1984).
14. P. Inarrea *et al., Eur. J. Pharmacol.* **105**, 309 (1984).
15. C.M. Chesney, D.D. Pifer, and K.M. Huch, *Platelet-Activating Factor and Structurally Related Ether Lipid*, J. Benveniste and B. Arnoux, Eds. (Elsevier, Amsterdam, 1983), pp. 177-184.
16. S.-B. Hwang, *J. Biol. Chem.* **263**, 3225 (1988).
17. S.-B. Hwang *et al., Biochemistry* **22**, 4756 (1983).
18. H. Homma, A. Tokumura, and D.J. Hanahan, *J. Biol. Chem.* **262**, 10582 (1987).
19. D.R. Janero, B. Burghardt, and C. Burghardt, *Thromb. Res.* **50**, 789 (1988).
20. L. Tahraoui *et al., Mol. Pharmacol.* **34**, 145 (1988).
21. F.H. Valone and E.J. Goetz, *Immunology* **48**, 141 (1983).
22. J.T. O'Flaherty *et al., J. Clin. Invest.* **78**, 381 (1986).
23. D.S. Ng and K. Wong, *Biochem. Biophys. Res. Commun.* **155**, 311 (1988).
24. V. Prpic *et al., J. Cell Biol.* **107**, 363 (1988).
25. W. Chao *et al., J. Biol. Chem.* **264**, 13591 (1989).
26. S.-B. Hwang, M.-H. Lam, and T.Y. Shen, *Biochem. Biophys. Res. Commun.* **128**, 972 (1985).
27. S.-B. Hwang, *Arch. Biochem. Biophys.* **257**, 339 (1987).
28. M.T. Domingo *et al., Biochem. Biophys. Res. Commun.* **151**, 730 (1988).
29. Ibid. **160**, 250 (1989).
30. A. Thierry *et al., Eur. J. Pharmacol.* **163**, 97 (1989).

31. S.-B. Hwang *et al., Lab. Invest.* **52,** 617 (1985).
32. S.-B. Hwang *et al., Eur. J. Pharmacol.* **120,** 33 (1986).
33. S.-B. Hwang and M.-H. Lam, *Biochem. Pharmacol.* **35,** 4511 (1986).
34. T.Y. Shen *et al., Platelet-Activating Factor and Related Lipid Mediators,* F. Snyder, Ed. (Plenum Press, New York, 1987), pp. 153-190.
35. R.N. Saunders and D.A. Handley, *Ann. Rev. Pharmacol. Toxicol.* **27,** 237 (1987).
36. K.H. Weber and H.O. Heuer, *Med. Res. Rev.* **9,** 181 (1989).
37. S.-B. Hwang *et al., J. Biol. Chem.* **260,** 15639 (1985).
38. Z. Terashita, Y. Imura, and K. Nishikawa, *Biochem. Pharmacol.* **43,** 1491 (1985).
39. T. Biftu *et al., J. Med. Chem.* **29,** 1919 (1986).
40. M.M. Ponpipom *et al., J. Med. Chem.* **30,** 136 (1987).
41. S.-B. Hwang, M.-H. Lam, and M.N. Chang, *J. Biol. Chem.* **261,** 13720 (1986).
42. C. Robaut *et al., Biochem. Pharmacol.* **36,** 3221 (1987).
43. O. Marquis, C. Robaut, and I. Cavero, *J. Pharmacol. Exp. Ther.* **244,** 709 (1988).
44. S.-B. Hwang, M.-H. Lam, and A.H.-M. Hsu, *Mol. Pharmacol.* **35,** 48 (1989).
45. S.-B. Hwang, *J. Lipid Med.*, in press.
46. D. Nunez, R. Kumar, and D.J. Hanahan, *Arch. Biochem. Biophys.* **272,** 466 (1989).
47. S.-B. Hwang, M.-H. Lam, and S.S. Pong, *J. Biol. Chem.* **261,** 532 (1986).
48. M.D. Houslay *et al., Biochem. J.* **238,** 109 (1986).
49. P.M. Lad, C.V. Olson, and I.S. Grewal, *Biochem. Biophys. Res. Commun.* **129,** 632 (1985).
50. D.L. Bratton *et al., Biochim. Biophys. Acta* **443,** 211 (1988).
51. D.L. Bratton *et al., Biochim. Biophys. Acta* **941,** 76 (1988).

3

PAF Interactions with Polymorphonuclear Neutrophils

Joseph T. O'Flaherty
Department of Medicine
Section on Infectious Diseases
Wake Forest University
Medical Center
Winston-Salem, NC
27103

Platelet-activating factor (PAF) has an uniquely broad spectrum of *in vitro* and *in vivo* bioactions. It activates polymorphonuclear neutrophils by binding with specific plasmalemmal receptors which induce N-proteins to activate phospholipases A_2 and C. The cells thus form inositol phosphates, diacylglycerol, lysophospholipids, and arachidonic acid. Inositol phosphates raise cytosolic Ca^{2+} levels; diacylglycerol activates protein kinase C (PKC) by a Ca^{2+}-enhanced reaction; lysophospholipids are converted to their *sn*-2 acetyl analogues (PAF-challenged cells acetylate alkyl ether lyso phosphocholine to make fresh PAF); and arachidonate is oxygenated to leukotrienes, hydroxyicosatetraenoates, and prostaglandins. The products interact to mediate function. PKC phosphorylates response-regulating proteins; leukotriene B_4, and endogenous PAF bind with their respective receptors to augment excitatory signal production; and 5-hydroxicosatetraenoate may up-regulate PAF receptors. PAF-stimulated neutrophils are also hyperresponsive to other agonists. This "primed" state may reflect fundamental changes in the cell's capacity to activate PKC. PAF treated cells soon become desensitized to PAF. This effect appears due to activated PKC, which down-regulates PAF receptors, and/or certain prostaglandins, which inhibit neutrophil responses to PAF.

Introduction

In 1971, Siraganian and Osler[1] observed that specific allergen induced sensitized leukocytes release a platelet-activating factor (PAF). Subsequent studies have shown that PAF likewise activates polymorphonuclear neutrophils (PMN);[2,3] eosinophils;[4,5] basophils;[6] lymphocytes;[7-10] monocytes;[11-13] vascular endothelium;[14-16] macrophages;[17-20] exocrine pancreas;[21] parotid acini;[21] neurocytes;[22] smooth muscle;[23-25] and cells from skin,[28,29] heart,[30] liver,[31,32] and kidney.[33-35] As indicated elsewhere in this volume, PAF is also implicated in diverse allergic, inflammatory, immunological, cardiovascular, and pulmonary reactions. Thus, PAF influences an extraordinarily wide range of cell types *in vitro* and has a multiplicity of pathological effects *in vivo*. The mechanisms by which PAF stimulates its many targets, then, are of interest. I review this area with respect to PMN. These cells are prototypical responders to PAF; knowledge gained with them is pertinent to PAF effects on the larger community of PAF-sensitive tissues.

As currently viewed, PMN responses to PAF pivot upon certain secondary messengers that form within seconds of PAF challenge and then proceed to regulate function. Accordingly, I focus on the production and activities of these transductional signals. But first, I discuss some physiochemical properties of PAF that influence its bioactions.

Structural and Physiochemical Considerations

PAF is an *sn*-glycero-3-phosphatidylcholine (GPC) containing any one of several long-chain fatty alkyl ether residues at *sn*-1 plus an acetate ester at *sn*-2. Among all natural molecular species of PAF, 1-hexadecyl-2-acetyl-GPC demonstrates the greatest potency; species with *sn*-1 alkyl ethers longer than 17 or shorter than 15 carbon units are less active in various bioassay systems.[36-38] Analogues lacking an *sn*-2 substituent or containing *sn*-2 alkyl, alkyl ether, or long-chain (that is, > 3 carbon units) acyl ester residues show decreased potency;[2,37] and phospholipids with an *sn*-1 fatty acid or *sn*-3 polar head group other than phosphocholine (for example, 1-alkyl-2-acetyl-phosphatidylethanolamine) have relatively little PMN-stimulating effects.[37] Finally, the inverted stereoisomer of PAF is 3 orders of magnitude weaker than natural PAF.[38] Interestingly, 1-hexadecyl-2-acetyl-GPC is the major molecular species of PAF produced by stimulated PMN.[39] PMN thus preferentially form optimally bioactive PAF.

PAF dissolves poorly in aqueous media. At > 1μM, it forms micelles,[40] disrupts membrane bilayers as well as whole cells,[41,42] and therefore has chaotropic, detergent-like properties. Albumin absorbs PAF to enhance its aqueous solubility while decreasing its apparent toxicity.[43] Albumin also promotes the presentation of PAF to cells. Hence, most bioassays use PAF as an albumin-absorbed mixture. In these instances, PAF bioavailability is protein-variant and reflects the lipid's partitioning between albumin, cells, and hydrophobic sinks.[44] This complicates interpretations of PAF binding and functional assays; the reported effects of PAF are very much dependent upon the conditions of measurement. Indeed, PAF bioactivities may at times be due to occult cell injury. Fortunately, PAF stimulates many cell types at concentrations much lower than 1μM, and the inverted PAF stereoisomer can be used to detect nonspecific responses. Finally, most cells and biological fluids have an acetylhydrolase that converts PAF to its bioinactive *sn*-2-lyso analogue. Acetylhydrolase-insensitive analogues (for example, 1-hexadecyl-2-*N*-methycarbamyl-GPC) are useful for probing the role of metabolism in PAF bioactions.[45]

PAF Receptors

Specific receptors transduce PAF-induced responses.[45-54] As seen in Table 1, studies report that PMN have either high- or high- plus low-affinity PAF receptors of widely varying affinities and cellular densities. These discrepancies reflect at least six problems in measuring PAF binding. First, PMN and their membranes rapidly convert

Table 1
Parameters of PAF Receptor Binding to PMN Reported by Various Laboratories[1]

			Receptor Affinity Type				
			—High—		—Low—		
Analogue	Temp (°C)	Method	K_d(nM)	R_t	K_d(nM)	R_t	Ref.
1-[^{3}H]alkyl-2-acetyl-GPC	37	F	0.1	5 x 10^6/cell	U	U	47
1-[^{3}H]alkyl-2-acetyl-GPC	22	F	4.5	3 x 10^4/cell	U	U	48
1-[^{3}H]hexadecyl-2-acetyl-GPC	4	F	0.2	1 x 10^3/cell	200	2 x 10^5/cell	49
1-[^{3}H]hexadecyl-2-acetyl-GPC	4	F	0.2	0.02pmol/mg	500	5pmol/mg	49
1-[^{3}H]aklyl-2-acetyl-GPC	24	F	0.6	0.4pmol/mg	0	0	50
1-[^{3}H]aklyl-2-acetyl-GPC	0	F	0.4	0.2pmol/mg	500	NR	51
1-[^{3}H]aklyl-2-acetyl-GPC	20	F	3.5	5 x 10^4/cell	0	0	52
1-[^{3}H]hexadecyl-2-acetyl-GPC	4	S	1.0	1.2 x 10^4/cell	700	9 x 10^5/cell	53
1-[^{3}H]hexadecyl-2-C-GPC[2]	4	F	1.1	2 x 10^3/cell	1200	2.4 x 10^5/cell	45
[^{3}H]52770	20	F	4.2	4 x 10^4/cell	0	0	52
[^{3}H]WEB 2086	25	F	16.3	4 x 10^4/cell	0	0	54

[1]PAF analogue binding to PMN was conducted at indicated temperatures using filtration (F) or sedimentation (S) methods to isotate cellular ligand. Results are reported as birding affinity (K_d) and quantity of receptors (R_t) in number per PMN or picomoles per mg of membrane protein. The latter result is given for those studies that examined ligand binding to isolated membrane preparations. Low affinity binding sites were not found (0) or unsaturable (U), as indicated.
[2]C stands for *N*-methycarbamyl.
NR = not reported.

PAF to inactive metabolites. Metabolism occurs at 37°C but not 4°C[47,55] and thus distorts studies done at physiological temperatures. Second, the washing of PMN removes PAF from its receptors.[53] Filtration assays, which commonly use washing techniques, thus give lower estimates of PAF binding than assays that avoid cell washing (for example, centrifugation of PMN through silicone oil). Third, commercially available [^{3}H]PAF is a mixture of molecular species,[56] some of which have relatively poor receptor binding properties. Such mixtures may underestimate binding relative to radiolabeled preparations of pure 1-[^{3}H]hexadecyl-2-acetyl-GPC. Fourth, many cations,[51] nucleotides,[50,51] proteins,[44] and lipids[57] have large effects on PMN binding of PAF. The binding assays from Table 1 differ in the content of these agents. Fifth, binding assays often employ low (for example, < 20nM) concentrations of PAF. This hinders detection of low-affinity (K_d > 100nM) binding sites. Sixth, some PAF analogues, such as the antagonists given in Table 1, tag different sets of PAF receptors other than those tagged by [^{3}H]PAF.[51,52] Clearly, then, assessments of PAF receptors are best considered as operational estimates that are limited to a defined set of assay conditions. Despite these difficulties, however, virtually all studies agree that: (1) PMN contain high-affinity PAF receptors; (2) these receptors transmit PAF bioactions; and (3) antagonists that block PAF receptor binding inhibit all PMN responses to PAF.[45,51,52,54,55,58-61]

PAF receptors (high- and low-affinity) localize to PMN plasma membrane. As judged using fractionation techniques and filtration binding assays, PMN have no extraplasmalemmal PAF specific binding sites, at least in a form that is both accessible to exogenous ligand and trapped with GF/C filters.[49] As indicated elsewhere in this volume, however, Hwang has shown that PMN may indeed have inaccessible PAF receptors. Studies with proteases[55] and sulfhydryl reagents[62] suggest that PAF receptors have peptide moieties on the PMN surface and a critical thio group(s) that is (are) needed for ligand binding. PAF receptors, as surmised using computer modeling of PAF analogue potencies, may have a central hydrophobic pocket (for insertion of the alkyl chain residue) embedded below two opposing zones of positive charge (for interaction with the PAF polar head group). PAF receptors thus resemble "earmuffs."[63]

Receptor Linkages With N-Proteins

Plasmalemmal receptors frequently cycle from an uncoupled, isolated, low-affinity state to a high-affinity state that is ligand bound and coupled to N-proteins. N-proteins are heterotrimers composed of α, β, and γ subunits. α Subunits from N_i- or N_o-proteins activate phospholipases C and, perhaps, phospholipases A_2, whereas α subunits from N_s-proteins activate adenyl cyclase. The ternary complex of ligand, receptor, and N-protein transduces function. The complex's N-proteins exchange their guanosine diphosphate (GDP) for cytosolic guanosine triphosphaste (GTP). This triggers N-protein dissembling, liberation of α subunits, and α-subunit-induced activation of phospholipases C and A_2 as well as adenyl cyclase. The original receptor, now freed from its N-protein, reverts to a low-affinity state. Soon, however, α subunits hydrolyze their GTP to GDP, assemble with β and γ subunits, and associate again with receptors to reform a high-affinity receptor/N-protein complex.[64]

PMN receptors for PAF appear to follow the reaction scheme just outlined. Thus, treatment of PMN membranes with PAF stimulates a GTPase;[50,65] GTP causes PMN membranes to lose high-affinity PAF binding sites, presumably by triggering receptor/N-protein dissociation;[49] and PMN challenged with PAF show evidence of phospholipases C and A_2 activation, that is, they cleave phosphatidylinositol phosphates at *sn*-3 [66-71] and release arachidonic acid from the *sn*-2 position of various phospholipids.[72-79] Moreover, pertussis toxin, which selectively blocks activation of N_i-and N_o-proteins, inhibits PAF-induced phospholipid metabolism.[66-70,79-82] Finally, PMN contain various N-proteins,[83-88] some of which have N_i/N_o-protein characteristics but are insensitive to pertussis toxin.[87] PMN also have N_s-protein activities, and PAF causes PMN to raise cyclic AMP.[74] This last effect, however, may proceed through a Ca^{2+}-dependent but N_s-protein-independent route.[89] The role of N_s-proteins in PAF-induced PMN function has not been extensively reported (see References 68 and 82). In any event, pertussis toxin-insensitive bioactions of PAF[67] may be mediated by pertussis toxin-insensitive N-proteins.

Ca^{2+} Transients

PAF-challenged PMN raise cytosolic Ca^{2+} levels.[53,66-71,80-82] Two distinct mechanisms appear responsible for this. First, inositol polyphosphates, formed by phospholipase C cleavage of phosphatidylinositols, release Ca^{2+} from subcellular storage pools.[70,71] The effect produces maximal rises in cytosolic Ca^{2+} within 30 seconds. Second, PAF causes extracellular Ca^{2+} to enter PMN.[67,90,91] The latter effect apparently produces a more sustained (time = 1 to 3 minutes) rise in cytosolic Ca^{2+}.[70,72] Ca^{2+} influx may be due to the opening of Ca^{2+} channels by: the initial (inositol phoshpate-induced) Ca^{2+} transient, that is, Ca^{2+}-sensitive Ca^{2+} channels;[90] N-proteins, that is, N-protein-linked Ca^{2+} channels;[92-94] inositol phoshpate metabolites, that is, inositol polyphosphate-activated Ca^{2+} channels;[95] or a postulated direct interaction with ligand-bound PAF receptors, that is, receptor-linked Ca^{2+} channels.[67,69]

Protein Kinase C Activation

Indirect studies indicate that PAF induces PMN to form diacylglycerol from phosphatidylinositols[66-71,95] (see also the studies on macrophages in Reference 19) by an N_i/N_o-protein-dependent, phospholipase C-mediated cleavage reaction. Based on recent studies,[96-104] it seems likely that PMN also form diacylglycerol from phosphatidylcholine *via* a second, N-protein-dependent pathway.[100-102] In any event, product diacyglycerol activates protein kinase C (PKC) by a Ca^{2+}-enhanced binding reaction.[53,105-107] PKC mobilization induced by PAF is maximal within 15 seconds and endures for 20 minutes before reversing.[105] Recent studies suggest that the early phase (time < 1 minute) of PKC mobilization results from rises in cytosolic Ca^{2+}, whereas later phases (times > 1 minute) may be triggered by some Ca^{2+}-independent event.[108]

Phospholipid Metabolism

PAF causes PMN to cleave *sn*-2-arachidonate from resident phospholipids.[72,77,109] Phospholipase A_2 catalyzes this reaction. The enzyme might be activated by N-proteins,[110] rises in cytosolic Ca^{2+},[111-113] or PKC.[114-116] The 1-alkyl- and 1-acyl-*sn*-2 lyso phospholipids formed by phospholipase A_2[39,117-121] are immediately acetylated to form their corresponding 1-radyl-2-acetyl phospholipids. Thus, PAF-treated PMN form, among other acetylated glycerolipids, PAF.[78,121,122] Acetyltransferase, which is responsible for acetylating lyso-PAF, and perhaps other lyso-phospholipids, may be activated by elevated cytosolic Ca^{2+},[123-127] cyclic AMP-dependent kinase,[128-130] or Ca^{2+}/calmodulin-dependent kinase.[131,132] In addition to this remodeling pathway, cells also form PAF by a *de novo* pathway.[133-136] As yet, however, there is no evidence that the latter pathway contributes to PAF-induced PAF formation, although activation of PKC is capable of stimulating *de novo* PAF synthesis in PMN.[135,136]

Arachidonic Acid Metabolism

PAF-stimulated PMN release arachidonate from their phospholipids *via* phospholipase A_2-mediated cleavage reactions. Furthermore, PMN taken from arachidonate-starved rats are poor producers of PAF, and this defect is reversed with exogenously administered arachidonate[137] (also see Reference 138). Thus, aracidonate release, phospholipid turnover, and PAF synthesis appear obligately coupled to each other. Released arachidonate is rapidly metabolized by 5-lipoxygenase to 5-hydroxyicosatetraenoate (5-HETE) and leukotriene (LT) B_4 as well as other oxygenated fatty acids.[72-76] PAF-stimulated PMN may also metabolize smaller amounts of arachidonate to prostaglandins.[72,76]

Excitatory Signals

The events just described produce excited PMN that are enriched with various bioactive signals. We know relatively little, however, about the ensuing steps that culminate in function. Rather, we have gross observations that implicate a given signal in one or another PMN response to PAF. For example, Ca^{2+} depletion, Ca^{2+} antagonists, PKC blockers, and anti-lipoxygenases partially inhibit PAF-induced PMN degranulation.[72,105,135-142] Presumably, then, Ca^{2+}, PKC, and hydroxylated arachidonate metabolites work together in mediating exocytosis of granule-bound enzymes. Nevertheless, this and similar interpretations are complicated by findings that: PAF causes PKC mobilization and stimulates synthesis of PAF, LTB_4, and 5-HETE; LTB_4 triggers PAF synthesis and PKC mobilization;[105,121] many of these agents raise cytosolic Ca^{2+};[53,66-71,82] rises in cytosolic Ca^{2+} trigger

PAF, LTB_4, and 5-HETE synthesis;[109,117,119,123-127] and activated PKC can influence all of these events.[135,136]

Thus, the many signals formed in PAF-challenged PMN can trigger their own as well as each others' formation. They may also have positive feedback effects. For example, 5-HETE operates by a stereospecific mechanism in PMN to increase PAF-induced degranulation, PKC mobilization, and phospholipid turnover.[105,121,141-144] By itself, however, the HETE has none of these actions and, moreover, has no such effects on the bioactions of, for example, chemotactic peptide agonists. Our preliminary studies suggest that 5-HETE causes PMN to increase the number of high-affinity receptors available to PAF. It seems at least possible, then, that the 5-HETE formed in PAF challenged PMN operates to promote PAF receptor availability and thereby enhances the ultimate functional response. Finally, a signal may operate in a response-specific fashion: endogenously formed LTB_4 appears essential for aggregation but not degranulation responses to PAF,[73,142,145] and extracellular Ca^{2+} is required for PAF-induced oxidative metabolism yet is completely irrelevant to the PMN priming actions (see following text) of PAF.[146,147] Thus, it seems that PAF causes a burst of excitatory signal formation. The signals may proceed to trigger each others formation, up-regulate PAF receptors, and cooperate in diverse, highly specific ways to elicit a particular functional response.

PAF as an Excitatory Signal

PMN produce PAF in response to not only PAF but also to various other stimuli.[115,117-122,127,135-137] This PAF may itself be an endogenous mediator of the rapid responses of PMN to various perturbants.[2,3,141] It also may be involved in PMN priming. PAF-pretreated PMN exhibit increased responsiveness to most other stimuli as judged in various assay systems.[2,146,148-151] This effect is independent of Ca^{2+} and arachidonic acid metabolism[146,148-150] but requires PAF receptors and PAF receptor/N-protein interactions.[146,149] The effect may involve PKC mobilization.[107] In any case, lipopolysaccharide,[150] tumor necrosis factor,[151] colony stimulating factor,[152,153] and chemotactic peptides[154,155] likewise prime PMN. Each of these agents, under cell-priming conditions, causes PMN to form PAF.[39,121,122,156-159] Studies with lipopolysaccharide suggest that this, and perhaps the other listed agents, may prime PMN by inducing endogenous PAF formation.[150]

Inhibitory Events

Many events work to antagonize excitatory signals: (1) activated components of N-proteins hydrolyze their GTP to GDP and then reassemble; (2) inositol phosphates and diacylglycerol are metabolically inactivated; (3) LTB_4 is oxidized to less active metabolites that are released from the cell;[161] 5-HETE is incorporated into phospholipids as well as oxidized and released extracellularly as 5,20-dihydroxyicosatetraenoate;[161] and Ca^{2+} is pumped out of cytosol. In addition to these more general events, PMN have more specific means for limiting the actions of PAF. First, PMN metabolize PAF to bioinactive derivatives.[55,162] This metabolism appears receptor-independent, may be inducible with cell-stimulating levels of PAF, and operates rapidly as well as quantitatively.[55] PAF metabolism occurs in PMN plasma membrane. At this site, a cytosolic acetylhydrolase deacetylates PAF, and a membranous transacylase transfers a long-chain fatty acid from other phosphatidylcholines to the *sn*-2 position of the lyso-PAF intermediate. The final acylated product then moves to granules, perhaps in association with specific transfer proteins.[48,163] Such processing, although not controlling the abrupt responses of PMN to PAF,[53] may operate to limit the more sustained actions of PAF (for example, chemotaxis). Second, PAF-induced mobilization of PKC, in contrast to that produced by, for example, chemotactic peptides, reverses after 20 minutes.[105] Third, prostaglandins, which form in stimulated PMN, are potent inhibitors of PAF-induced PMN responses.[164] Finally, under many conditions, PAF receptors down-regulate within seconds of binding PAF. This effect is mimicked by various PKC activators and inhibited by PKC blockers.[52,165] Kinetic studies suggest that receptor down-regulation occurs with sufficient rapidity to explain the termination of responses and desensitization that occurs in PAF-challenged PMN.[165] Thus, PKC may be a critical stop signal in limiting the PMN-stimulating actions of PAF.

Summary and Conclusions

PAF stimulates PMN by binding to high-affinity plasmalemmal receptors. The receptors, when ligand-bound, functionalize N-proteins which, in turn, activate phospholipases. Phospholipase-mediated cleavage of resident phospholipids results in the formation of inositol phosphates, diacylglycerol, and various arachidonate metabolites. Inositol phosphates release storage Ca^{2+}, diacylglycerol activates PKC by a Ca^{2+}-enhanced reaction, and arachidonate metabolites like LTB and 5-HETE promote further metabolism of phospholipids. LTB_4 accomplishes this by binding to its own specific, N-protein-linked receptors,[160] whereas 5-HETE may act by up-regulating PAF receptors. The pro-excitatory mediators proceed to orchestrate events that produce function. Concurrently, PMN metabolically inactivate plasmalemmal PAF and transfer the final product to intracellular granules. Finally, the PMN enter a less active, desensitized state. Phospholipid turnover ceases, bioactive lipid products are converted to less active metabolites, cytosolic Ca^{2+} returns to prestimulatory levels, PAF receptors down-regulate, and PKC mobilization reverses. PKC may cause PAF receptor down-regulation, and certain arachidonate metabolites, that is, prostaglandins, may further dampen cellular responsiveness. PAF-treated and desensitized PMN, however, are not totally inactive. They exhibit greatly increased reactivity to agonists other than PAF. Moreover, they gradually regain full sensitivity to further PAF challenge over the ensuing one to two hours.[165] The PAF-PMN interaction, then, seems best described as involving receptor-induced production of excitatory as well as inhibitory mediators that act in concert to elicit and then terminate function. The PAF receptor itself appears to be an important target of these mediators. For example, 5-HETE may up-regulate, whereas PKC may down-regulate, PAF receptor availability and/or function. Ultimately, however, mediator levels decline, and PMN as well as PAF receptors[165] regain their prestimulatory responsiveness. The entire PAF stimulatory event, then, can be described as a cyclical passage of PMN through resting, excitatory, and PAF-desensitized (but otherwise primed) phases, with an ultimate return to the normal resting state.

Acknowledgments

Our original work site in this manuscript was supported by National Institutes of Health Grants HL-27799 and HL-26257. Thanks to Jean Kimbrell for help in document preparation.

References

1. R.P. Siraganian and A.G. Osler, *J. Immunol.* **106**, 1244 (1971).
2. J.T. O'Flaherty *et al., J. Immunol.* **127**, 731 (1981).
3. J.O. Shaw *et al., J. Immunol.* **127**, 1250 (1981).
4. A.J. Wardlaw *et al., J. Clin. Invest.* **78**, 1701 (1986).
5. G. Kimani, M.G. Tonnesen, and P.M. Henson, *J. Immunol.* **140**, 3161 (1988).
6. B.S. Bochner *et al., J. Clin. Invest.* **81**, 1355 (1988).
7. P. Patrignani *et al., Biochem. Biophys. Res. Commun.* **148**, 802 (1987).
8. A. Dulioust *et al., J. Immunol.* **140**, 240 (1988).
9. M. Rola-Pleszczynski *et al., J. Immunol.* **140**, 3547 (1988).
10. E. Vivier *et al., Eur. J. Immunol.* **18**, 425 (1988).
11. T. Yasaka, L.A. Boxer, and R.L. Baehner, *J. Immunol.* **128**, 1939 (1982).
12. B. Czarnetzki, *Clin. Exp. Immunol.* **54**, 486 (1983).
13. G. Barzaghi, H.M. Sarau, and S. Mong, *J. Pharmacol. Exp. Ther.* **248**, 559 (1989).
14. J.J. Emeis and C. Kluft, *Blood* **66**, 86 (1985).
15. F. Bussolino *et al., J. Immunol.* **139**, 2439 (1987)
16. G.Y. Grigorian and U.S. Ryan, *Circ. Res.* **61**, 389 (1987).
17. H.-P. Hartung, *FEBS Lett.* **160**, 209 (1983).
18. H. Hayashi *et al., J. Biochem.* **97**, 1255 (1985).
19. R.J. Uhing *et al., J. Biol. Chem.* **264**, 9224 (1989).
20. V. Pripic *et al., J. Cell Biol.* **107**, 363 (1988).
21. H.-D. Soling, H. Eibi, and W. Fest, *Eur. J. Biochem.* **144**, 65 (1984).
22. E. Kornecki and Y.H. Ehrlich, *Science* **240**, 1792 (1988).
23. N.P. Stimler and J.T. O'Flaherty, *Am. J. Pathol.* **113**, 75 (1983).
24. J. Nishihira *et al., Lipids* **19**, 907 (1984).
25. U.S. Schwertschlag and A.R. Whorton, *J. Biol. Chem.* **263**, 13791 (1988).

26. M. Kester, R. Kumar, and D.J. Hanahan, *Biochim. Biophys. Acta* **888**, 306 (1986).
27. M.K. Ferguson, H.K. Shahinian, and F. Michelassi, *J. Surg. Res.* **44**, 172 (1988).
28. H. Kawaguchi and H. Yasuda, *FEBS Lett.* **176**, 93 (1984).
29. G.J. Fisher *et al., Biochem. Biophys. Res. Commun.* **163**, 1344 (1989).
30. J.L. Kenzora *et al., J. Clin. Invest.* **74**, 1193 (1984).
31. S.D. Shukla *et al., J. Biol. Chem.* **258**, 10212 (1983).
32. F. Mendlovic, S. Corvera, and J.A. Garcia-Sainz, *Biochem. Biophys. Res. Commun.* **123**, 507 (1984).
33. J.V. Bonventre, P.C. Weber, and J.H. Gronich, *Am. J. Physiol.* **254**, F87 (1988).
34. J. Pfeilschifter, A. Kurtz, and C. Bauer, *Biochem. Biophys. Res. Commun.* **127**, 903 (1985).
35. D. Schlondorff and R. Neuwirth, *Am. J. Physiol.* **251**, F1 (1986).
36. G.F.E. Scherer, *Biochem. Biophys. Res. Commun.* **133**, 1160 (1985).
37. J.T. O'Flaherty *et al., Res. Commun. Chem. Pathol. Pharmacol.* **39**, 291 (1983).
38. R.L. Wykle *et al., Biochem. Biophys. Res. Commun.* **100**, 1651 (1981).
39. H.W. Mueller, J.T. O'Flaherty, and R.L. Wykle, *J. Biol. Chem.* **259**, 14554 (1984).
40. W. Kramp *et al., Chem. Phys. Lipids* **35**, 49 (1984).
41. D.R. Hoffman, J. Hajdu, and F. Snyder, *Blood* **63**, 545 (1984).
42. D.B. Sawyer and O.S. Andersen, *Biochim. Biophys. Acta* **987**, 129 (1989).
43. J.C. Ludwig *et al., Arch. Biochem. Biophys.* **241**, 337 (1985).
44. R.N. Pinckard, in *Platelet-Activating Factor and Diseases*, K. Saito and D.J. Hanahan, Eds. (International Medical Publishers, Minato-ku, Tokyo 106, Japan, 1989), pp. 37-50.
45. J.T. O'Flaherty *et al., Biochem. Biophys. Res. Commun.* **147**, 18 (1987).
46. G. Camussi *et al., Panminerva Med.* **22**, 1 (1980).
47. F.H. Valone and E.J. Goetzl, *Immunology* **48**, 141 (1983).
48. F. Bussolino, C. Tetta, and G. Camussi, *Agents Actions* **15**, 15 (1984).
49. J.T. O'Flaherty *et al., J. Clin. Invest.* **78**, 381 (1986).
50. D.S. Ng and K. Wong, *Biochem. Biophys. Res. Commun.* **141**, 353 (1986).
51. S.-B. Hwang, *J. Biol. Chem.* **263**, 3225 (1988).
52. 0. Marquis, C. Robaut, I. Cavero, *J. Pharmacol. Exp. Ther.* **244**, 709 (1988).
53. J.T. O'Flaherty, D.P. Jacobson, and J.F. Redman, *J. Biol. Chem.* **264**, 6836 (1989).
54. G. Dent *et al., FEBS Lett.* **244**, 365 (1989).
55. J.T. O'Flaherty *et al., FEBS Lett.* **250**, 341 (1989).
56. H.W. Mueller, J.T. O'Flaherty, R.L. Wykle, *J. Biol. Chem.* **259**, 14554 (1984).
57. J.T. O'Flaherty and D. Jacobson, *Biochem. Biophys. Res. Commun.* **163**, 1456 (1989).
58. T.Y. Shen *et al., Proc. Natl. Acad. Sci. USA* **82**, 672 (1985).
59. J. Casals-Stenzel, G. Muacevic, and K.-H. Weber, *J. Pharmacol. Exp. Ther.* **241**, 974 (1987).
60. S.-B. Hwang *et al., J. Pharmacol. Exp. Ther.* **246**, 534 (1988).
61. G.A. Zimmerman *et al., J. Cell Biol.* **110**, 529 (1990).
62. D.S. Ng and K. Wong, *Eur. J. Pharmacol.* **154**, 47 (1988).
63. G. Dive *et al., J. Lipid Mediators* **1**, 201 (1989).
64. E.J. Neer and D.E. Clapham, *Nature (London)* **333**, 129 (1988).
65. T. Matsumoto *et al., Biochem. Biophys. Res. Commun.* **143**, 489 (1987).
66. P.H. Naccache *et al., Biochem. Biophys. Res. Commun.* **130**, 677 (1985).
67. P.H. NacGache *et al., J. Leukocyte Biol.* **40**, 533 (1986).
68. E.L. Becker, *Fed. Proc.* **45**, 2148 (1986).

69. E.L. Becker *et al., Fed. Proc.* **45,** 2151 (1986).
70. M.W. Verghese *et al., J. Immunol.* **138,** 4374 (1987).
71. A.G. Rossi, R.M. McMillan, and D.E. MacIntyre, *Agents Actions* **24,** 272 (1988).
72. F.H. Chilton *et al., J. Biol. Chem.* **257,** 5402 (1982).
73. A.H. Lin, D.R. Morton, and R.R. Gorman, *J. Clin. Invest.* **70,** 1058 (1982).
74. R.R. Gorman *et al., Leukotriene Res.* **12,** 57 (1983).
75. L.M. Ingraham *et al. , Blood* **59,** 1259 (1982).
76. M.-L. Dahl, *Int. Arch. Allergy Appl. Immunol.* **76,** 145 (1985).
77. J.-S. Tou, *Lipids* **22,** 333 (1987).
78. T.W. Doebber and M.S. Wu, *Proc. Natl. Acad. Sci. USA* **84,** 7757 (1987).
79. W. Tao, T.F.P. Molski, R.I. Sha'afi, *Biochem. J.* **257,** 633 (1989).
80. P.M. Lad, C.V. Olson, I.S. Grewal, *Biochem. Biophys. Res. Commun.* **129,** 632 (1985).
81. P.M. Lad *et al., Proc. Natl. Acad. Sci. USA* **82,** 8643 (1985).
82. J.T. O'Flaherty, D. Jacobson, and J. Redman, *J. Immunol.* **140,** 4323 (1988).
83. P. Gierschik *et al., J. Biol. Chem.* **261,** 8058 (1986).
84. B.D. Gomperts, M.M. Barrowman, and S. Cockcroft, *Fed. Proc.* **45,** 2156 (1986).
85. M. Oinuma, T. Katada, and M. Ui, *J. Biol. Chem.* **262,** 8347 (1987).
86. G.M. Bokoch, C.A. Parkos, and S.M. Mumby, *J. Biol. Chem.* **263,** 16744 (1988).
87. I. Fuse and H.-H. Tai, *Biochem. Biophys. Res. Commun.* **146,** 659 (1987).
88. P. Goldsmith *et al., J. Biol. Chem.* **262,** 14683 (1987).
89. M.W. Verghese *et al., J. Biol. Chem.* **260,** 6769 (1985).
90. V. von Tscharner *et al., Nature (London)* **324,** 369 (1986).
91. J.E. Merritt, R. Jacob, and T.J. Hallam, *J. Biol. Chem.* **264,** 1522 (1989).
92. B.D. Gomperts, *Nature (London)* **306,** 64 (1983).
93. R.H. Scott and A.C. Dolphin, *Nature (London)* **330,** 760 (1987).
94. F. DiVirgilio *et al., J. Biol. Chem.* **262,** 4574 (1987).
95. S.B. Dillon *et al., J. Biol. Chem.* **262,** 11546 (1987).
96. L.W. Daniel, M. Waite, and R.L. Wykle, *J. Biol. Chem.* **261,** 9128 (1986).
97. J.M. Besterman, V. Duronio, and P. Cuatrecasas, *Proc. Natl. Acad. Sci. USA* **83,** 6785 (1986).
98. J. Reibman *et al., J. Biol. Chem.* **263,** 6322 (1988).
99. L.G. Rider, R.W. Dougherty, and J.E. Niedel, *J. Immunol.* **140,** 200 (1988).
100. D.E. Agwu *et al., J. Biol. Chem.* **264,** 1405 (1989).
101. J.-K. Pai *et al., J. Biol. Chem.* **263,** 12472 (1988).
102. A.P. Truett III, R. Snyderman, and J.J. Murray, *Biochem. J.* **260,** 909 (1989).
103. J. H. Exton, *J. Biol. Chem.* **265,** 1 (1990).
104. C.D. Smith, C.C. Cox, and R. Snyderman, *Science* **232,** 97 (1986).
105. J.T. O'Flaherty and J. Nishihira, *J. Immunol.* **138,** 1889 (1987).
106. J.C. Gay and E.S. Stitt, *Blood* **71,** 159 (1988).
107. J.C. Gay and E. S. Stitt, *J. Cell. Physiol.* **137,** 439 (1988).
108. J.T. O'Flaherty *et al., J. Biol. Chem.*, in press.
109. H.W. Mueller, J.T. O'Flaherty, and R.L. Wykle, *J. Biol. Chem.* **258,** 6213 (1983).
110. S. Nakashima *et al., Arch. Biochem. Biophys.* **261,** 375 (1988).
111. B.J. Bormann *et al., Proc. Natl. Acad. Sci. USA* **81,** 767 (1984).
112. T. Matsumoto, W. Tao, R.I. Sha'afi, *Biochem. J.* **250,** 343 (1988).
113. J. Balsinde *et al., J. Biol. Chem.* **263,** 1929 (1988).
114. W.C. Liles, K.E. Meier, and W.R. Henderson, *J. Immunol.* **138,** 3396 (1987).

115. T.M. McIntyre *et al., J. Biol. Chem.* **262**, 15370 (1987).
116. S.L. Reinhold *et al., J. Biol. Chem.* **264**, 21652 (1989).
117. C.L. Swendsen *et al., Biochem. Biophys. Res. Commun.* **113**, 72 (1983).
118. F.H. Chilton *et al., J. Biol. Chem.* **259**, 12014 (1984).
119. F.H. Chilton and T.R. Connell, *J. Biol. Chem.* **263**, 5260 (1988).
120. M. Cluzel, B. J. Undem, F. H. Chilton, *J. Immunol.* **143**, 3659 (1989).
121. T.G. Tessner, J.T. O'Flaherty, and R.L. Wykle, *J. Biol. Chem.* **264**, 4794 (1989).
122. J. Gomez-Cambronero *et al., J. Biol. Chem.* **264**, 21699 (1989).
123. R.L. Wykle, B. Malone, and F. Snyder, *J. Biol. Chem.* **255**, 10256 (1980).
124. T.-C. Lee *et al., Biochem. Biophys. Res. Commun.* **105**, 1303 (1982).
125. J. Gomez-Cambronero *et al., Biochem. J.* **219**, 419 (1984).
126. J. Gomez-Cambronero *et al., Biochim. Biophys. Acta* **845**, 511 (1985).
127. M.M. Billah *et al., J. Biol. Chem.* **261**, 5824 (1986).
128. J. Gomez-Cambronero *et al., Biochim. Biophys. Acta* **845**, 516 (1985).
129. J. Gomez-Cambronero *et al., Biochem. J.* **245**, 893
130. M.L. Nieto, S. Velasco, and M. Sanchez Crespo, *J. Biol. Chem.* **263**, 4607 (1988).
131. M.M. Billah and M.I. Siegel, *Biochem. Biophys. Res. Commun.* **118**, 629 (1984).
132. C. Domenech, E.M. De Domenech, and H.-D. Soling, *J. Biol. Chem.* **262**, 5671 (1987).
133. T.-C. Lee, B. Malone, and F. Snyder, *J. Biol. Chem.* **261**, 5373 (1986).
134. Ibid., **263**, 1755 (1988).
135. M.L. Nieto, S. Velasco, and M. Sanchez Crespo, *J. Biol. Chem.* **263**, 2217 (1988).
136. S. Leyravaud *et al., J. Immunol.* **143**, 245 (1989).
137. C.S. Ramesha and W.C. Pickett, *J. Biol. Chem.* **261**, 7592 (1986).
138. E. Remy *et al., Biochim. Biophys. Acta* **1005**, 87 (1989).
139. J.T. O'Flaherty *et al., Am. J. Pathol.* **105**, 107 (1981).
140. J.T. O'Flaherty *et al., Am. J. Pathol.* **105**, 264 (1981).
141. J.T. O'Flaherty, *J. Cell Physiol.* **122**, 229 (1985).
142. Ibid., **125**, 192 (1985).
143. A.G. Rossi, M.J. Thomas, and J.T. O'Flaherty, *FEBS Lett.* **240**, 163 (1988).
144. A.G. Rossi and J.T. O'Flaherty, *Lipids* in press.
145. I. Moodley and A. Siuttle, *Prostaglandins* **33**, 253 (1987).
146. G.M. Vercellotti *et al., Blood* **71**, 1100 (1988).
147. K.T. Hartiala *et al., Biochem. Biophys. Res. Commun.* **144**, 794 (1987).
148. J.C. Gay *et al., Blood* **67**, 931 (1986).
149. M. Shalit, G.A. Dabiri, and F.S. Southwick, *Blood* **70**, 1921 (1987).
150. L.A. Guthrie *et al., J. Exp. Med.* **160**, 1656 (1984).
151. Y.H. Atkinson *et al., J. Clin. Invest.* **81**, 759 (1988).
152. R.H. Weisbart *et al., Blood* **69**, 18 (1987).
153. A. Yuo *et al., Blood* **74**, 2144 (1989).
154. A.B. Karnad *et al., Blood* **74**, 2519 (1989).
155. D.E. Van Epps and M.L. Garcia, *J. Clin. Invest.* **66**, 167 (1980).
156. G.S. Worthen *et al., J. Immunol.* **140**, 3553 (1988).
157. G. Camussi *et al., J. Exp. Med.* **166**, 1390 (1987).
158. U. Wirthmueller, A.L. De Weck, and C.A. Dahinden, *J. Immunol.* **142**, 3213 (1989).
159. J.H. Sisson *et al., J. Immunol.* **138**, 3918 (1987).
160. J.T. O'Flaherty, J.F. Redman, and D.P. Jacobson, *J. Cell. Physiol.* **142**, 299 (1990).
161. J.T. O'Flaherty *et al., J. Immunol.* **137**, 3277 (1986).

162. F.H. Chilton *et al., J. Biol. Chem.* **258**, 6357 (1983).
163. J.B. Banks *et al., Biochim. Biophys. Acta* **961**, 48 (1988).
164. A.G. Rossi and J.T. O'Flaherty, *Prostaglandins* **37**, 641 (1989).
165. J.T. O'Flaherty, D. Jacobson, and J. Redman, submitted.

4

Hetrazepines as PAF Antagonists

Hetrazepine platelet-activating factor (PAF) antagonists represent a potent group of substances that compete with high affinity for the PAF-binding site of both intracellular and cell surface membrane PAF receptors. Although related in structure to certain hypnotic drugs in clinical use, these antagonists have been found, both in animals and man, to be devoid of such effects on the central nervous system. Studies in animals have shown activity in models of (among other diseases) asthma, shock, cardiovascular disease, gastrointestinal disease, stroke, and transplantation rejection. Studies in man are under way.

correspondence
C.J. Meade
H.O. Heuer
Lung Pharmacology Group
Boehringer Ingelheim KG
D-6507 Ingelheim am Rhein
West Germany

Discovery of the PAF-Antagonistic Activity of Hetrazepines

Hetrazepines are defined as oxazepines (X=O), thiazepines (X=S), or diazepines (X=N), with two annelated five-membered heterocycles (see Figure 1).

Compounds of this type have been used for many years as hypnotic or anxiolytic drugs. In 1983, Jorge Casals-Stenzel, working in the laboratories of Boehringer Ingelheim in Germany, discovered that some of these drugs also had the ability to inhibit the action of platelet-activating factor (PAF), as measured by human platelet aggregation. The benzo-triazolo-1,4-diazepines triazolam and alprazolam, and especially the thieno-triazolo-1,4-diazepine brotizolam

Figure 1. A generalized molecular model of a hetrazepine compound.

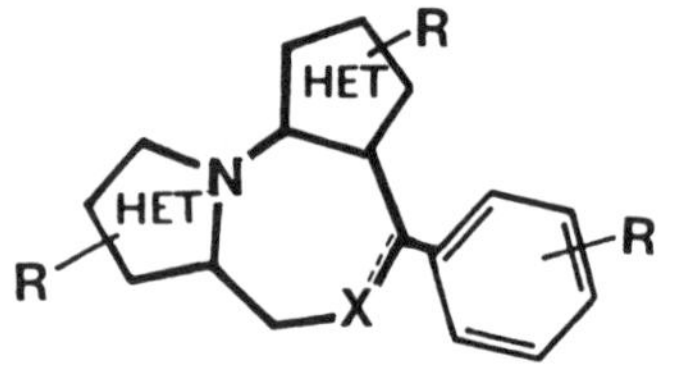

(Figure 2, Structure 1), were particularly active. These compounds not only had a pronounced ability to inhibit PAF-induced platelet aggregation but also had a specificity of action in that there was little or no effect on aggregation induced by adenosine diphosphate (ADP), adrenaline, collagen, serotonin, or arachidonic acid.[1,2]

At about the same time, but working independently, Kornecki and her group[3] in the United States also discovered that triazolam and alprazolam had PAF-antagonistic activity.

However, attempts to show that these triazolodiazepines also acted *in vivo* against the effects of PAF were difficult because the doses of drug required to inhibit PAF effects were high in relation to the doses required to produce sedation and hypnosis. Further progress required the dissociation of the PAF-antagonistic and central nervous system (CNS) effects of this class of compounds.

Dissociation of the PAF-Antagonistic and CNS Effects of Hetrazepines

Benzodiazepines exert their hypnotic and sedative effects through specific tissue receptors in the CNS. A second type of benzodiazepine receptor exists on peripheral tissues, including platelets.[4-6] Casals[7] found that RO 15-1788 and RO 5-4864, antagonists of the central and peripheral benzodiazepine receptors, respectively, not only did not themselves have significant activity against PAF-induced platelet aggregation; they also failed to block the PAF-antagonistic activity of brotizolam. Furthermore, in *in vivo* experiments in guinea pigs, a dose

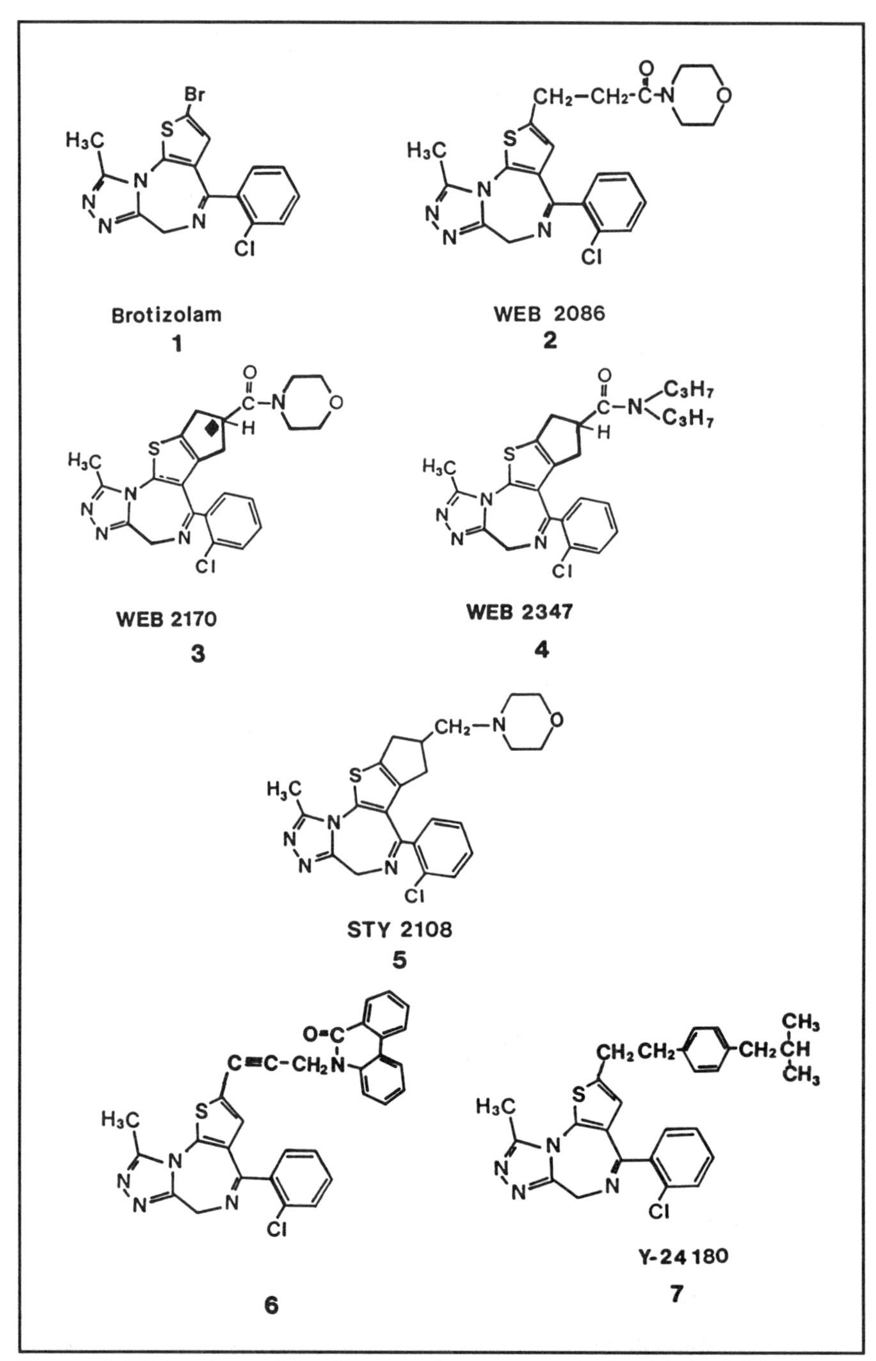

Figure 2. Molecular models of hetrazepine compounds that inhibit PAF.

of RO 15-1788 high enough to completely block the hypnotic and motor-depressant effects of brotizolam (or triazolam) not only did not inhibit PAF-induced bronchoconstriction and hypotension; it also did not prevent brotizolam and triazolam themselves from blocking these PAF effects.

Discovery of WEB 2086, WEB 2170, and Related Compounds

The pharmacological separation of the hypnotic and PAF-inhibitory activities of brotizolam stimulated an effort to modify the brotizolam molecule so that these activities were separated structurally. It became clear early that one key to doing this was substitution on the 2-position of the thiophene ring. Modification of the brotizolam molecule could reduce or even abolish binding to the central type benzodiazepine receptor while at the same time enhancing binding to the PAF receptor. In addition, by inserting hydrophilic groups in the 2-position, the water solubility of the molecules could be enhanced. This reduced lipophilicity was important for decreasing the ability of the molecules to cross the blood-brain barrier. It was the custom in Boehringer Ingelheim at the time this synthetic work began to identify compounds by the initial three letters of the name of the chemist who had first made the synthesis. The two thousand and eighty-sixth compound synthesized by Karl-Heinz Weber and co-workers[8,9] was not only a potent PAF-antagonist but also had no CNS effects and was water soluble. WEB 2086 (see Figure 2, Structure 2) was selected for development. It also soon became a valuable tool and standard in PAF research.

At Boehringer Ingelheim, studies on the modification of the side chain at position 2 on the thiophene ring were then extended by ring closure to give a new series of tetracyclic compounds that had both potent PAF-antagonistic activity and low affinity for the benzodiazepine receptor. WEB 2170 (see Figure 2, Structure 3), WEB 2347 (see Figure 2, Structure 4), and STY 2108 (see Figure 2, Structure 5) are examples of this series.[10-13]

Once the underlying principle of the importance of substitution at the 2-position of the thiophene ring had been defined, there was (and is!) no limit to the number of compounds that can be synthesized with substitutions in this position and with the potential of dissociated PAF-antagonistic and CNS effects. Study of these compounds does not add any superior insight into the basic findings that have already been described but does provide interesting examples to illustrate the fundamental principles. In Y-24180 (see Figure 2, Structure 7), the 2-position is substituted with an arylalkyl moiety (isobutyl phenylethyl).[14] Side chains consisting of aryl or bicyclic and tricyclic heterocyclic moieties, bound to the 2-position of the thiophene ring by a propynyl bridge (for example, Structure 6), also confer powerful PAF-antagonistic activity.[15] BN 50739 is an example of hetrazepine with a side chain including a methoxysubstituted benzyl group.[16]

It is not the aim of this review to go into details of the structure-activity relationships of the hetrazepine series, which in any case have already been described in some detail in previous publications[12,13] and the patent literature. The explanation of these structure-activity relationships has provided an interesting challenge for computer-modelers. Published modeling studies have attempted to draw analogies between the PAF molecule and not only WEB 2086 but also the structurally totally dissimilar PAF antagonist BN 52021.[17]

Receptor Binding Characteristics of WEB 2086, WEB 2170, WEB 2347, and Other Hetrazepines

WEB 2086 (and when studied WEB 2170 or WEB 2347) have proven, in all tissues so far studied, to act as potent, competitive antagonists, binding at the same site as the PAF molecule.

Studies on the receptor binding characteristics of these molecules have been facilitated by the availability of radiolabeled compounds (WEB 2086 is even available commercially, from New England Nuclear, Boston, MA).

Labeled WEB 2086 bound to human platelets with high affinity (K_D = 6.1nM) and unlabeled PAF competed for this binding. Similarily, unlabeled WEB 2086 competed for the binding of radiolabeled

PAF. Indirect measurements of WEB 2086 binding affinity, by measuring (^{3}H)PAF-binding in the presence of differing quantities of unlabeled WEB 2086 and then applying the Schild equation, gave similar orders of affinity as the direct measurements (K_B = 8.3nM). The K_i's for inhibition of binding of (^{3}H)PAF or (^{3}H)WEB 2086 by either of these substances unlabeled were similar. Also, within the limits of experimental error, the calculated number of binding sites for WEB 2086 per platelet (260) was the same as the calculated number of binding sites for PAF (240).[18]

Evidence that WEB 2086 functions by competition at the level of the PAF receptor also came from studies in which PAF-induced aggregation rather the (^{3}H)PAF binding was the study parameter. Dose-response curves for PAF were shifted to the right in a parallel fashion by increasing concentrations of WEB 2086, and the slope of the Schild plot [log antagonist concentration versus log (dose ratio - 1)] was not significantly different from one.[19]

Studies measuring receptor occupancy and inhibition of platelet aggregation by WEB 2086 suggested that at least 80 percent occupancy of PAF receptors by WEB 2086 was required to block the effect of PAF on platelet aggregation.[20]

Further evidence that the PAF receptor occupied by WEB 2086 is functional came from studies comparing the ability of different hetrazepines or hetrazepines and nonhetrazepines in their ability either to displace (^{3}H)PAF or inhibit platelet aggregation .[12,21,22] There was a correlation between these two events. Further, compared with nonhetrazepines, WEB 2086 has shown itself to be a particularily potent inhibitor of both PAF binding and PAF-induced platelet aggregation.[19,21] More recently described hetrazepines such as STY 2108 have been shown to be yet more active.[12]

Despite the high affinity of binding to the PAF receptor, the affinity of binding to the central diazepine receptor [assessed by (^{3}H)flunitrazepam binding] was negligible.[13] The ability of WEB 2086 to inhibit PAF-induced platelet aggregation was not affected by the central benzodiazepine receptor ligand RO 15-1788 or the peripheral benzodiazepine ligand RO 5-4864.[23] Hetrazepines with ring closure at the 2-position on the thiophene ring showed even lower binding to the central benzodiazepine receptor than WEB 2086.[13] The high degree of specificity of WEB 2086 for the PAF receptor was further demonstrated by studies in which the potent inhibition of human platelet aggregation induced by PAF (IC_{50} = 0.17μM) contrasted with the absence of any effect, except at very high (> 100μM) concentration, on aggregation induced by ADP, collagen, serotonin, or arachidonic acid. Unlike the PAF antagonists RO 19-3704 and CV-3988, WEB 2086 did not significantly inhibit aggregation induced by adrenaline.[8,24]

The exquisite stereoselectivity of hetrazepine binding is illustrated by WEB 2170, which has a chiral center (◆ in the structural formula). The (–) optical enantiomer not only bound with considerably greater affinity (K_i = 14μM) but was also considerably more potent in inhibiting platelet aggregation (IC_{50} = 0.35μM) than the (+) isomer (K_i = 660μM, IC_{50} = 8.79μM).[25]

The binding kinetics of (^{3}H)PAF and (^{3}H)WEB 2086 to human platelets are compatible with the existence of a single type of high-affinity PAF receptor, occupancy of which does not influence the affinity of other receptors.[18] This need not mean, however, that all platelet functions are inhibited by WEB 2086 at similar concentrations, and in guinea pig platelets, PAF-induced chemoluminescence [a measure of adenosine triphosphate (ATP) formation] required less WEB 2086 for inhibition than PAF-induced aggregation.[26]

Binding sites for WEB 2086 have been reported in a variety of other cell types including neutrophils[27] and eosinophils.[28] This correlates with the ability of WEB 2086 to inhibit PAF-induced changes in function: in neutrophils, inhibition of aggregation,[8] β-glucuronidase release,[27] and chemotaxis;[29] in eosinophils, inhibition of degranulation,[30-32] superoxide production,[31,32] calcium influx,[32,33] and chemotaxis.[29]

Inhibition of PAF-induced changes in cell function has also been described in macrophages[34] as well as in the monocytic cell line U 937.[35] In comparison with other nonhetrazepine PAF antagonists such as BN 52021 or L-652,731, WEB 2086 appeared more potent as shown by the concentration of drug required for the inhibition of PAF-induced elevation of intracellular calcium in U937 cells or by inhibition of PAF-induced prostacyclin production by peritoneal macrophages. The calculated pA_2 values not only for WEB 2086 but also for the other PAF antagonists tested were higher for inhibition of macrophage prostacyclin production than for inhibition of platelet or polymorphonuclear leukocyte aggregation, a finding that has been interpreted in terms of receptor subtypes. The existence of PAF receptor subtypes has also been postulated to explain why, for example, in a comparison of WEB 2086 and another PAF antagonist, CV-6209 (which is of a totally different structural type), WEB 2086 was more efficient in blocking neutrophil chemotaxis, but CV-6209 was more efficient in blocking eosinophil chemotaxis.[29] It could also explain why in human eosinophils higher concentrations of PAF are required to stimulate, and higher concentrations of WEB 2086 are required to block, PAF-induced superoxide production in comparison with eosinophil peroxidase release and increase in intracellular calcium.[32] The concept of PAF-receptor subtypes has received support from other work such as that described by Hwang *et al.* (see Chapter 2 in this volume) in this volume, but direct evidence for their existence (as might be provided, for example, by isolation and gene cloning) is not at present available. Other possibilities exist to explain the results with WEB 2086. For example, occupation of the PAF receptor has, in several cell types, been shown to stimulate inositol polyphosphate production, an effect blocked by WEB 2086.[36] The products of polyphosphoinositide breakdown in turn regulate kinase activity and intracellular calcium release. Differences in the threshold levels of kinase activity or calcium concentration required for the control of different cell processes could explain why inhibition of different processes requires different levels of receptor occupancy by WEB 2086, without requiring the postulation of receptor subtypes. Alternatively, results could also be explained if a single PAF receptor existed in more than one conformational form, with accompanying differences in receptor affinity. There is good evidence that such conformational and affinity changes, regulated by ionic environment (for example, Ca^{2+}, Mg^{++} concentration), can occur.[37] In the example quoted earlier[32] in which different eosinophil functions were inhibited at different WEB 2086 concentrations, different ionic conditions were present in the test systems because PAF-induced superoxide production was a magnesium ion-dependent process and PAF-induced peroxidase release was calcium ion-dependent.

It is likely that WEB 2086 is able to cross cell membranes and block intracellular as well as extracellular PAF receptors. The evidence for this comes from studies in which cells were stimulated with agents under conditions in which PAF production was increased but there was very little release of this phospholipid to the extracellular medium. For example, endothelial cells were stimulated with bradykinin[38] or macrophages with formyl-methionyl-leucyl-phenylalanine.[39] At the same time, prostaglandin I_2 was released with a time course that closely followed that of PAF production. Inhibition of PGI_2 production by WEB 2086 (as well as by a structurally unrelated PAF antagonist, CV-6209) has been used as evidence that intracellular PAF has a messenger role in signal transduction and can only be explained if WEB 2086 also blocks intracellular receptors. The inhibition of the intracellular PAF receptor on endothelial cells may have important effects on the ability of inflammatory cells such as neutrophils to adhere to the endothelial cell surface.

In contrast to certain other PAF antagonists, particularly those with structures more closely related to the PAF molecule such as CV-3988,[40,41] there is no evidence that hetrazepine PAF antagonists have either nonspecific membrane effects or PAF-agonistic activity. Although the degradative enzyme PAF-acetylhydrolase also has a binding site for PAF, and CV-3988 can inhibit this enzyme,[42] patients treated with WEB 2086 have shown no reduction in serum PAF-acetylhydrolase activity (C.J. Meade, unpublished results, 1989).

The identity of the PAF- and WEB 2086-binding sites has now been so well established that WEB 2086 can be considered a tool of choice for identifying PAF receptor sites. In this respect it has several advantages over PAF itself because of the presence in many tissues of abundant low-affinity but probably nonfunctional PAF binding sites, and the readiness with which PAF can be broken down by PAF acetylhydrolase.[43] This probably explains, for example, the wide divergence in number and affinity of PAF receptor sites reported, on the basis of studies with a PAF ligand, in human platelets and leukocytes.[44] In addition, WEB 2086 has advantages over other PAF antagonists for enumerating PAF-receptor sites in its high degree of specificity. By comparison, several other inhibitors bind multiple sites (for example, SCH 37370 binds both histamine and PAF receptors), or even though these inhibitors competitively inhibit PAF binding, they seem to do so by a mechanism other than by direct interaction with the PAF receptor site; for example, 52770 RP blocked PAF-induced platelet aggregation but inhibited binding to the platelet only of itself and not of radiolabeled PAF or WEB 2086.[18]

WEB 2086 is suitable for identifying PAF receptors not only on cell lines or in membrane preparations but also in tissue sections. An example of such an application, which is of importance in view of the clinical testing of WEB 2086 under way in pulmonary diseases, is the identification of high-affinity WEB 2086 binding sites in human[45] (as well as guinea pig[45,46]) lung membranes. Autoradiography of human lung sections shows binding sites widely distributed over peripheral airways, vessels, and parenchyma but not on the segmental airways.[45] The ability of WEB 2086 to block PAF effects on isolated human bronchus[47] or *in vivo* in lung fuction in human volunteers[48] suggests that these binding sites are functional.

The specificity of WEB 2086 and the detailed characterization of the WEB 2086 binding site on platelets also enables this compound to be used in the *in vitro* assay of PAF from biological samples as a means of checking that platelet-aggregatory activity is indeed due to PAF and not due to other factors able to aggregate platelets (for example, thromboxanes).[49]

Effects of Hetrazepines in Animal Models

Almost all of the pharmacological effects of PAF in experimental animals can be reversed with WEB 2086 or WEB 2170; all major routes of administration (by mouth, by aerosol, or by injection intravenously or at other sites) are appropriate.

Effects on the Lung

Inhibition of the Effects of PAF Itself

The effects of WEB 2086 as well as other hetrazepines on the lung have been studied not only in whole animals but also in the isolated organ and on individual lung tissues.

Probably the most easily monitored effect of PAF on the lungs is bronchoconstriction, and it is for this reason that this parameter has been widely studied, even though clinically the role of PAF antagonists is more likely to be in the control of the inflammatory correlates of asthma rather than in the inhibition of bronchoconstriction. In guinea pigs, the mechanism of PAF-induced bronchoconstriction is dependent on the route of administration. Bronchoconstriction induced by intravenously administered PAF is platelet dependent, whereas bronchoconstriction induced by PAF given via the trachea is not.[51] Bronchoconstriction induced by PAF given by either route can be inhibited by oral, intravenous, and inhalative WEB 2086,[8,52,53] as well as (where tested) by WEB 2170,[54,55] WEB 2347,[56] or STY 2108.[54,55] The ability of WEB 2086 aerosol to block not only the bronchoconstriction induced by PAF infusion but also the concomitant fall in systemic blood pressure[8] suggests that there is good systemic absorption of WEB 2086 from the lung.

Information on localization of the site of action of PAF (and of WEB 2086) has been provided by studies with isolated tissues. The effects of PAF on isolated segmental airways are notoriously difficult to reproduce, a finding that is perhaps understandable in the light of the absence of PAF receptors on some segmental airways.[45] In those human bronchi specimens able to respond by contraction to 0.1μM PAF, the effect was blocked by WEB 2086. PAF-induced hyperresponsiveness to histamine was also blocked.[47] Tracheal strips do not contract to PAF (rather, they are relaxed), but PAF could enhance the contractile reactivity to K^+, an effect abolished by 10nM WEB 2086.[57] Tested on guinea pig parenchymal strips, even low (0.05μM) concentrations of PAF produced a sustained contraction, an effect blocked in a concentration-dependent manner by WEB 2086 (IC_{50} = 1.6μM).[58] A second study on the same tissue but using higher (0.5μM) concentrations of PAF as stimulus confirmed this result, although the concentrations of WEB 2086 required for inhibition were also higher.[59]

The mechanism of PAF-induced bronchoconstriction has also been dissected using isolated lung preparations. In the guinea pig, 10 or 30ng PAF, perfused through the pulmonary artery, caused bronchoconstriction that could be blocked by simultaneous perfusion with 0.1μM WEB 2086.[52] Also, PAF-induced release of the bronchoconstricting agent, thromboxane, was blocked by WEB 2086 in a dose-dependent manner. However, thromboxane release (as well as bronchoconstriction) induced by arachidonic acid was unaffected, showing the specificity of WEB 2086 for PAF-induced effects. The perfusate in these studies was Krebs' solution, not blood, so that PAF receptors blocked by WEB 2086 and responsible for thromboxane production and bronchoconstriction were probably tissue receptors and not located on blood elements such as platelets. The lung tissue in this study came from normal, nonimmunized animals. When lungs from immunized animals were used, different results were obtained. In comparison with lungs from nonimmunized animals, lungs from actively sensitized guinea pigs are likely to be infiltrated with appreciable numbers of blood-derived cells. Where active inflammatory processes are present, there is also the possibility of re-routing of the blood supply away from inflamed areas. Whatever the explanation, it appears that WEB 2086 (like the structurally unrelated PAF antagonist BN 52021) was much less effective at blocking PAF-induced bronchoconstriction and release of mediators (histamine and leukotriene-like material) when tested on lungs from actively sensitized rather than unsensitized guinea pigs.[60] Nevertheless, even if the lungs came from sensitized animals, and bronchoconstriction and mediator release was unaffected, WEB 2086 was still able to suppress PAF-induced edema formation as measured by increased lung wet weight.

Studies of isolated rat lungs have also shown the ability of WEB 2086 to block parameters associated with pulmonary edema. In one study, WEB 2086 was shown to block, in a dose-related manner, the increase in wet-to-dry lung weight, pulmonary artery perfusion pressure, and bronchial infiltration pressure induced by a 20μg intraarterial PAF bolus.[61] A later study confirmed some of these findings; 10μM of WEB 2086 block PAF-induced increase in lung weight and substantially reduced the increase in perfusion pressure.[62] Because a 5-lipoxygenase inhibitor (CGS 8515) also substantially reduced these parameters, and PAF stimulated the release of leukotriene-like material into the perfusate, it is likely that PAF effects in this model are, at least in part, secondary to leukotriene release. Measured leukotriene release was blocked by WEB 2086.

Another approach to assessing pulmonary edema is measurement of leakage of plasma proteins labeled either with radioactive isotopes or dyes. In the lung, PAF-induced extravasation of either radiolabeled fibrinogen,[63] albumin, or the dyes fluorescein,[64] copper phthalocyanine,[65] or Evans blue[66] was completely blocked by WEB 2086. Both immediate and delayed (5-hour) effects of PAF were inhibited.[64] In one study,[66] WEB 2086 was compared with another PAF antagonist, the ginkgolide mixture BN 52063. WEB 2086 showed maximal and complete inhibitory action at 10μg/kg IV, although the dose of BN 52063 required to completely block extravasation was 500-fold higher. In another study, using histological localization of carbon particles as the method of quantitating extravasation,[67] not only was WEB 2086 active at blocking PAF-induced extravasation, but also specificity for PAF could be demonstrated by the failure of WEB 2086 to block histamine- or leukotriene D_4-induced microvascular leakage.

At the same time that PAF increases vessel permeability to proteins, it also leads to recruitment of proinflammatory cells from the vascular compartment into lung tissue. Eosinophils are a cell type whose infiltration into the lung is a particularly characteristic feature of asthma. It has been suggested that destruction of pulmonary epithelium by products of activated eosinophils underlies the bronchial hyperreactivity that is characteristic of this disease. The ability of PAF to stimulate *in vitro* eosinophil chemotaxis, and WEB 2086 to block this effect, is paralleled by *in vivo* observations that intravenous WEB 2086 (3mg/kg) caused appreciable reduction in the degree of eosinophil infiltration into the bronchial wall of guinea pigs six hours after inhalation of PAF.[68] At the same time, epithelial damage was decreased.

The PAF-induced infiltration of eosinophils into the bronchial walls is to a large extent platelet dependent.[68] PAF-induced accumulation of ^{111}In-labeled platelets in the thoracic region was also blocked by WEB 2086[8] as well as WEB 2170 and STY 2108.[54,55]

Two other effects of inhaled PAF are the formation of mucus plugs and decrease in tracheal mucus velocity associated with mucociliary dysfunction. These effects could be inhibited *in vivo* by WEB 2086.[68,69] They may be in part mediated via eosinophils, because *in vitro* the damaging effects of PAF on guinea pig ciliated tracheal rings were greatly enhanced if eosinophils were also present.[70]

The long-term effects of PAF (via inflammation, cell infiltration, epithelial damage) are associated with alterations in pulmonary function. Aerosolized PAF has been reported to induce not only an "immediate" bronchoconstriction, but also a "late-phase "(six to eight hours) response. However, neither in man nor in experimental animals do all subjects respond in this way. When present, the "late phase" response (like "early-phase" PAF-induced bronchoconstriction) was, in a sheep model, blocked by WEB 2086.[71]

PAF has also been reported to increase responsiveness to nonspecific bronchoconstrictors, although again the effects observed are often small, and not all subjects always respond. WEB 2086 (1mg/kg IV) blocked hyperreactivity to acetylcholine induced in guinea pigs by 600ng/kg/hour PAF.[72,73] WEB 2086, when given 30 minutes prior to aerosol PAF challenge, has also been found to block hyperreactivity to carbachol in sheep.[74]

Activity in Lung Disease Models

Models of anaphylaxis and bronchial asthma. Although PAF is a powerful bronchoconstrictor, it is likely that it provides only a small contribution to immediate bronchoconstriction induced by allergen or other stimuli in most animal models. Thus, studies have been published in which WEB 2086, at doses capable of inhibiting PAF-induced effects, failed to influence immediate allergen-induced bronchoconstriction in actively sensitized guinea pigs[52] or rabbits.[75] Those models in which it has been possible to demonstrate an effect of PAF antagonists often require special conditions such as coadministration of low-dose antihistamine[76-78] or antihistamine plus lipoxygenase inhibitor;[79,80] antigen challenge by the aerosol rather than the intravenous route;[78] or, if by the intravenous route, then by very high doses of antigen;[77] or use of particular sensitization procedures, for example, passive sensitization.[52,76]

Isolated organ and cell-culture techniques have helped provide insight into the possible mechanisms of inhibition by WEB 2086 and other hetrazepines of antigen-induced bronchoconstriction. As mentioned previously, PAF produces only modest and somewhat variable direct effects on bronchial smooth muscle, and it is likely that other mediators (particularly histamine in the "early phase," and thromboxane and leukotrienes later in the response) are more important as end-mediators of allergic bronchoconstriction. Partial inhibition of release of one or more of these mediators has been shown in isolated lung preparations. Production of leukotriene D_4-like material, from rat lungs challenged with immune complexes[81] or thromboxane and histamine from actively or passively sensitized guinea pig lungs challenged with antigen,[52] was partially blocked by WEB 2086. Attempts to inhibit histamine release from basophilic leukocytes of human atopic subjects have, however, been unsuccesful.[82] On isolated lung strips from sensitized animals, an activity of PAF antagonist hetrazepines such as alprazolam or WEB 2086 in blocking antigen-induced contraction could only be demonstrated in the presence of drug "cocktails," usually including an antihistamine and an inhibitor of leukotriene synthesis or activity.[52,79,80,83] Timing of effects was important. In rabbit peripheral lung strips, from animals sensitized since birth so that the allergic response included a major immunoglobulin E component, the early (first five minutes) effects of antigen were not blocked by WEB 2086 (for this, an antihistamine was necessary), but after 40 minutes, WEB 2086 was effective, although an antihistamine was not.[83]

As just described, both *in vivo* and *in vitro* studies suggest that in most animal models of "early phase" bronchoconstriction, histamine and perhaps other mediators such as leukotrienes play a more important role than PAF, so that WEB 2086 and other hetrazepines are of only limited effectiveness against immediate bronchoconstriction. However, clinically, "early-phase" asthmatic reactions are relatively easy to treat symptomatically, and it is for the therapy of late-phase reactions and bronchial hyperreactivity that the need for new drugs is most pressing. In animal models involving both an "early" and "late" bronchoconstriction response to antigen challenge, hetrazepine inhibition of the "late" response can be found even when the "mmediate" response is unaffected.

In a study with ovalbumin-sensitized, conscious guinea pigs, the "very immediate" (that is, first few minutes) effect of challenge with aerosolized antigen could not be influenced by WEB 2086 given one hour prior to challenge (0.5mg/kg p.o.), although some effect on the fall in specific airways conductance occurring at 2 and 24 hours was observed (mainly a reduction in the duration of response). The 17-hour and 72-hour "late" responses could be entirely blocked by WEB 2086 given six hours after challenge, an observation that suggests that it is PAF released *after* the early reaction is over that is more critical for the development of late responses.

In the same experiments, WEB 2086 given one hour prior to challenge was able to block the allergen-induced rise in eosinophils and neutrophils in the fluid collected by bronchoalveolar lavage at 17 hours post challenge, while WEB 2086 given six hours post challenge was able to block the allergen-induced rise in the number of eosinophils measured at 72 hours.[84] The number of neutrophils counted after 72 hours was not significantly affected by WEB 2086.

This finding agrees with the observed inhibition by WEB 2086 of eosinophil infiltration into the bronchial walls as assessed by histological examination six hours after intravenous administration of ovalbumin to passively sensitized guinea pigs.[68]

In a study using conscious sheep,[71] WEB 2086 (1mg/kg IV, given

20 minutes before challenge) substantially reduced the late (6 to 8 hours) response to inhaled *Ascaris suum*, although again the immediate response was not affected. Pretreatment with WEB 2086 could also reduce the hyperreactivity (measured two hours after *A. suum* inhalation) to a non-specific challenge agent, carbachol.[85]

However, in another study, WEB 2086, even at doses at which it inhibited PAF-induced bronchial hyperreactivity, was unable to prevent propranolol or indomethacin from causing hyperreactivity to histamine in ventilated, anesthetized guinea pigs.[73] In this respect, WEB 2086 was similar to the two other PAF-antagonists studied, CV-3988 and BN 52021.

To summarize this section, hetrazepine PAF antagonists such as WEB 2086 can, in appropriate models, be shown to block bronchoconstriction, eosinophil influx, and bronchial hyperreactivity, three important features of clinical asthma. However, for each of these processes, choice of model is critical for the demonstration of beneficial effects. Only experience in the clinic will show which models are more appropriate for clinical asthma.

Models of increased permeability of lung vessels and "shock lung." WEB 2086 was effective in blocking pulmonary microvascular leakage caused by bradykinin,[86] endotoxin,[87] or (in rats) anaphylactic shock[62] but was not effective in blocking the changes associated with the immediate anaphylactic response to antigen in guinea pigs.[66] Such a difference is not surprising in view of the known potency of bradykinin and endotoxin as activators of phospholipase A_2, whereas in immediate anaphylaxis in the guinea pig, histamine is believed to be the most important mediator.

WEB 2086 was effective in reducing endotoxin-induced lung microvascular leakage in rats (as measured by ^{125}I-albumin efflux) not only when given prophylactically but also when given therapeutically 30 or 90 minutes after endotoxin challenge.[87] In conscious sheep, WEB 2086 inhibited the *late* increase of lymph flow and lymph protein clearance following intravenous administration of endotoxin, although endotoxin-induced changes in pulmonary artery pressure or lung mechanics were unaffected.[88]

Lung damage is an important factor contributing to mortality in septicemia and endotoxic shock. In conjunction with the rat studies previously cited,[87] WEB 2086, even when given two hours after administration of toxin, was found to significantly reduce the lethal effects of high-dose *Salmonella enteritidis* endotoxin. Improved survival has, in addition, been demonstrated following peroral or intravenous pretreatment with WEB 2086 in rats injected with potentially lethal doses of *Escherichia coli* toxin.[89]

Orally administered WEB 2170 or WEB 2347 could also exert a protective effect in mice injected with endotoxin; in these experiments, the animals were also injected with either propranolol or primed with tumor necrosis factor to enhance the effect of the endotoxin.[90,91] Set against these studies, however, are other studies in which hetrazepine PAF antagonists failed to reduce the lethal effect of endotoxin. It is perhaps not surprising that not all studies have yielded the same result, because the causes of death following endotoxin treatment are likely to be different according to the species selected, route and dose of endotoxin administration, and so forth, and not all the pharmacological or toxicological effects of endotoxin can be reversed by PAF antagonists.

Models of pulmonary venoconstriction and pulmonary artery hypertension. The anoxia that is, in the clinic, associated with emphysema and other chronic obstructive pulmonary diseases can cause pulmonary venoconstriction, pulmonary artery hypertension, and eventually right heart failure. PAF perfused into the pulmonary artery in a rat isolated lung preparation caused a rise in pulmonary artery pressure, an effect blocked by WEB 2086. WEB 2086 also blocked in a dose-dependent manner the increase in perfusion pressure when the partial oxygen pressure of the blood used to perfuse the lung was reduced. This was an activity not shared by another PAF antagonist, BN 52021.[92] However, when WEB 2086 was tested *in vivo* in pigs, doses sufficient to block PAF effects had no beneficial effect on hypoxia-induced pulmonary hypertension.[93]

Enhanced pulmonary venoconstriction also plays a role in the toxic action of polycations. Very low doses of PAF, which alone were not injurious, enhanced the toxic effects of protamine on isolated rat lungs; enhancement of pulmonary microvascular pressure suggests that pulmonary venoconstriction was involved. WEB 2086 blocked this effect.[94]

Gastrointestinal System

PAF is a potent ulcerogenic agent, and it has been proposed that release of endogenous PAF during septicemic shock may contribute to the damage to both gastric and gut mucosa that is characteristic of this condition. PAF also has profound effects on both gastric and gut motility and has been suggested to be involved in certain types of colic, including postoperative ileus.[95-98] PAF has also been implicated as stimulating ion fluxes across the colonic mucosa. Passiveefflux of water in association with such fluxes could have a pro-diarrheal effect.[100]

WEB 2086 blocked gastric ulceration produced by administration of a PAF bolus.[96] It also blocked the reduction of intestinal transit velocity in mice following intravenous PAF injection.[97,98] The effects of PAF in this system are enhanced by pretreatment with endotoxin or tumor necrosis factor; WEB 2086 could also block this priming reaction. In ponies implanted with gastrointestinal electrodes and strain gauges, the effects of endotoxin on gut contraction were reduced by administration of WEB 2086.[99] However, the effects on ion fluxes across the rat colonic mucosa (measured by transepithelial potential difference or short circuit current measured under voltage clamp conditions) were not affected by WEB 2086 or by the other PAF receptor antagonists tested,[100] which suggests that either this particular effect was not receptor mediated or that the PAF receptor involved was of a type different from most other PAF receptors.

In a rat model of chronic colitis induced by intracolonic administration of trinitrobenzene sulfonic acid, daily treatment with WEB 2086 or WEB 2170 resulted in a significant reduction of the colonic damage.[101]

Cardiovascular System

Modification of PAF Effects on the Cardiovascular System

Injecting PAF produces a profound but reversible systemic hypotension. This effect is produced even in rats, whose platelets are only poorly aggregated by PAF and so is probably a platelet-independent phenomenon.[102] Inhibition of PAF-induced systemic hypotension by WEB 2086, WEB 2170, or other PAF antagonist hetrazepines has been shown in rats,[8,54,56,103] guinea pigs,[8,54,55] and the monkey *Macaca fascicularis*.[104] Specificity of inhibition was shown in the last named study by the inability of WEB 2086 to block histamine-induced hypotension.

Because PAF still induced, and WEB 2086 still blocked, systemic hypotension in pithed and vagotomized rats, peripheral mechanisms are likely to be at least partially responsible for mediating the hemodynamic effects of PAF and WEB 2086.[103] One possible peripheral mechanism is an increase in the diameter of arterioles.

Macroscopic observations on the exteriorized mesenteric circulation of anesthetized rats demonstrated an increase in mean arteriolar diameter after PAF perfusion or local PAF application. This effect was blocked by WEB 2086.[105] In a variant of this study, the mesentery was treated with the vasoconstrictor noradrenaline, and the time required for interruption of blood flow was measured. PAF increased this latency period, and WEB 2086 blocked this effect.[106]

Ability of WEB 2086 to inhibit PAF-induced peripheral vasodilation has also been shown in an autoperfused rat hind limb preparation. Increasing concentrations of WEB 2086 shifted the dose-response curves for the vasodilating effects of bolus doses of PAF to the right. In contrast, as evidence of specificity, the vasodilating effect of acetylcholine was not affected.[107]

The effect of PAF on the heart is to cause coronary vasoconstriction and a decrease in myocardial contractility. In the perfused, isolated guinea pig heart, WEB 2086 blocked PAF- induced vasoconstriction.[108,109] It also (at concentrations between 0.03 and 1μM) antagonized in a dose-dependent fashion the sustained increase of perfusion pressure and decrease of cardiac developed tension elicited by a bolus injection of 50pmol PAF.[110] This effect was specific, because changes produced by leukotriene C_4 or the thromboxane mimetic U44069 were not blocked. The effects of PAF may be indirect, because PAF induced the release of thromboxane B_2 and leukotriene-like material; both of these latter agents are potent coronary vasoconstrictors,[109] and WEB 2086 specifically blocked PAF-induced leukotriene and thromboxane release.[109,110]

By contrast to the coronary vasoconstriction, effects of PAF on myocardial contractility (as expressed by the first differential of left-ventricular pressure, LV *dP/dt.*) are probably not even in part thromboxane mediated, because the thromboxan antagonist BM 13505 could not influence PAF-induced reduction of myocardial contractility under conditions in which WEB 2086 was active.[109]

The results of these studies with isolated hearts have been corroborated by *in vivo* studies using the monkey *M. fascicularis,* in which the LV *dP/dt.* was dose-dependently reduced by PAF, and WEB 2086 (0.22μmol/kg IV) attenuated this effect.[104]

Constriction of the coronary blood vessels is a factor contributing to cardiac ischemia. In guinea pigs, WEB 2086 completely blocked PAF-induced signs of ischemia [derived from the electrocardiogram (ECG)]. The drug also blocked PAF-induced cardiac arrhythmias[111,112] and was able to block PAF-induced worsening of arrhythmias caused by other agents such as ouabain and (in open chest dogs) experimental deprivation of coronary blood flow.[112] These PAF effects were probably mediated via thromboxanes and, to a lesser extent, lipoxygenase products, because the thromboxane antagonist BM 13177 and, to a limited extent, lipoxygenase inhibitors such as esculetin, were also active in this system .

Activity in Models of Cardiac Ischemia

In models of ischemia, rather than just PAF-induced cardiac damage, WEB 2086 was able to improve some indices of disease severity but not others. Experimental coronary stenosis in conscious dogs was, following reperfusion, associated with an increase of coronary vascular resistance in the stunned areas. This effect was blocked by administration of WEB 2086.[113] WEB 2086 also reduced the extent of myocardial necrosis for any given flow reduction and selectively improved endocardial flow in the perfused area.[114] However, in rabbits, WEB 2086 (even at doses sufficient to block PAF-induced blood pressure fall) did not block the effects of temporary coronary artery occlusion on either cardiac edema or neutrophil accumulation.[115]

The beneficial effects of WEB 2086 in some models of cardiac ischemia suggest a possible role in the treatment of myocardial infarction or unstable angina pectoris.

Modification of PAF Effects on the Blood-Cell Picture

Infusion of PAF into experimental animals induces thrombocytopenia and either leukocytosis or leukopenia. These effects show a slightly different profile of pharmacological inhibition from *in vitro* platelet or leukocyte aggregation, which suggests that aggregatory substances released from the blood vessel wall also play a role in the *in vivo* phenomenon. This concept is further supported by the ability of PAF to induce thrombocytopenia in rats, although rat platelets in autologous plasma are relatively unresponsive to PAF. WEB 2086 blocked PAF-induced thrombocytopenia in rats.[116] In guinea pigs, not only PAF-induced reduction in blood-platelet count but also accumulation of ^{111}In-labeled platelets in the thoracic region could be blocked by WEB 2086, WEB 2170, or STY 2108.[8,52,54,55]

Activity in Models of Shock

In addition to changes in the lungs and gastrointestinal tract , as previously described, septicemic shock is also associated with a fall in blood pressure, thrombocytopenia, and abnormalities in the pattern or number of blood leukocytes. Injection of endotoxin into experimental animals provides a model for these changes. Both in human septicemia and experimental endotoxin shock, there is evidence for an association of shock with elevated levels of PAF.[117,118]

The hypotensive action of endotoxin in rats could be reduced or abolished by treatment with WEB 2086, WEB 2170, or STY 2108. Not only were these substances active when given before toxin, they could also act therapeutically.[54,89-91,119] Intravenous injection at the nadir of endotoxin-induced hypotension (approximately six minutes after toxin injection) dose-dependently reversed the endotoxin effect. Unlike many other pharmacological agents that are also active in preventing endotoxin-induced hypotension (for example, adrenaline) the PAF antagonists alone did not raise blood pressure.

Reversal of both the cardiac and the vasodilatory effects of endotoxin may contribute to the efficacy of the PAF antagonists. Vasodilation during systemic endotoxin shock has been investigated in an autoperfused rat hind limb preparation. WEB 2086 (10μM in the perfusate plus 1mg/kg administered intravenously) partially reversed an endotoxin-induced fall in hind limb perfusion pressure.[107] WEB 2086 was also able to block the endotoxin-induced increase in arteriolar diameter measured in a rat exteriorized mesentery preparation. The doses required were equivalent to those required to block PAF effects.[105]

The studies in rats just described used relatively high doses of endotoxin (1 to 20mg/kg) given over short time periods. In another model in miniature pigs, much lower doses of toxin (2μg/kg/hour lipopolysaccharide from *Salmonella abortus equii*) were given over a longer time period. In this model, WEB 2086 (10mg/kg/hour) failed to prevent systemic hypotension, although the accompanying pulmonary hypertension was reduced. Also, WEB 2086 was able to retard the toxin-induced leukopenia and to partially prevent deterioration of both gas exchange and the mechanical properties of the lung.[120]

In addition to its beneficial action in endotoxin shock, WEB 2086 can also improve shock states resulting from trauma or hemorrhage. When exposed to Noble-Collip drum trauma, anesthetized rats treated with 0.5mg/kg WEB 2086 showed higher survival rates and were better able to maintain their mean arterial blood pressure than were untreated controls. Other indicators of traumatic shock such as increased plasma levels of lysosomal hydrolase, cathepsin D, and free amino-nitrogen compounds were also improved by WEB 2086 treatment. Release into the bloodstream of the cardiotoxic peptide myocardial depressant factor was blunted.[121]

Together, the beneficial effects of hetrazepine PAF antagonists on both cardiovascular and pulmonary consequences of shock suggest they may be useful as therapeutic agents, provided they can be given early enough. Enthusiasm for the hetrazepine PAF antagonists in this context must, however, be tempered with caution, because other agents (for example, corticosteroids) able to cause excellent protection in animal models have proved only of marginal value when tested in the clinical setting in individuals with established shock.

Central Nervous System

Even though it is unlikely that the water-soluble PAF-antagonistic hetrazepines such as WEB 2086 cross the blood-brain barrier to any significant extent, they may have a role in the treatment of diseases in which the blood-brain barrier is compromised, such as stroke. The ability of the hetrazepine BN 50739 to protect rabbits from some of the effects of lumbar spinal cord ischemia is described by Yue and colleagues elsewhere in this volume. In a rat middle cerebral artery occlusion model, treatment with WEB 2086 beginning either one half hour before or one hour after occlusion was able to significantly reduce both cortical and total infarct volume, as assessed planimetrically on brain sections taken 48 hours post occlusion.[122]

The hypnotic actions of brotizolam occur at lower doses than the PAF-antagonistic actions. Nevertheless, it is possible that PAF-antagonistic activity may contribute to the overall pharmacological profile and may provide advantages for brotizolam (especially high-dose brotizolam) over hypnotics not having this property. Brotizolam (10mg/kg IP) has been studied in a bilateral carotid artery occlusion model using mongolian gerbils. Given therapeutically one hour after declamping, the drug improved postischemic recovery, as assessed by improvement in the mitochondrial respiratory control ratio.[123]

Models of Inflammation and Graft Rejection

Apart from the studies in the lung mentioned earlier, WEB 2086 has been tested for effects on inflammation (edema, cell migration) at a number of other sites. It has been shown to block PAF-induced edema in the skin,[8,40,66] bladder, and esophagus[66] and to inhibit paw swelling[124] and entry of fluid into the pleural cavity following local PAF injection in mice.[124,125] It was also effective in blocking both the exudation and cell-population changes occurring after PAF injection into the pleural cavity of rats, although another PAF antagonist, 48740 RP, was only active against the exudation and not the changes in pleural leukocyte count.[125]

WEB 2086 also showed some activity against inflammation induced by agents other than PAF. As might be expected in view of the multiplicity of potential mediators of inflammation, the effects were generally only partial and very dependent on the nature of the system used, species, inflammation-inducing agent, and time of measurement of response. WEB 2086 attenuated the inflammatory edema induced by intraplantar injection of a crude extract of coelenterate tentacles.[126] It was active against zymosan-induced pleurisy in rats but not against that induced by carrageenan, which supports an important role for PAF only in inflammation induced by the former agent. This finding agrees with the significant reduction of the pleural response to zymosan but not carrageenan, following PAF-induced autodesensitization.[125]

Effects on nonimmunological (for example, inflammatory) components of the transplantation rejection reaction may underly the observed prolongation by WEB 2170 of survival of heterotopic cardiac allografts in rats also treated with low-dose cyclosporine.[127]

Table 1
Comparison of Different PAF Antagonists *in vitro* and *in vivo*

	K_I(nM)	IC_{50}(μM)	ED_{50} (mg/kg IV)		
	Platelets				
Substance	Binding (a)	Aggregation (b)	BC/GP (c)	BP/GP (d)	BP/Rat (e)
CV-3988	180	46.2	1.9	1.8	0.21
RO 19.3704	3	0.83	0.09	0.04	0.2
Ginkgolide B	470	0.82	1.2	1.1	6.4
L 652.731	130	1.3	2.9	2.1	0.6
48.740 RP	20000	48.9	21.0	15.0	3.2
52770 RP	22	7.6	1.3	1.4	6.7
WEB 2086	15	0.17	0.016	0.015	0.5

Methods/models are (a) binding of ^{3}H-PAF to washed human platelets, (b) 5 x 10^{-8} M PAF-induced human platelet aggregation (platelet-rich plasma), (c) bronchoconstriction and (d) hypotension in guinea pigs produced by an infusion of PAF at 30ng/(kg x minutes) PAF, (e) hypotension produced by 30ng/(kg x minutes) PAF infused into rats (from Reference 53).

An Overview of the Effects of Hetrazepine PAF Antagonists in Animal Models: Potencies, Half-Lives, and Choice of Treatment Regimen in New Studies

Undoubtedly WEB 2086, WEB 2170, and other PAF-antagonistic hetrazepines will be investigated in many more animal models not only in the disease areas previously described but also in other disease areas (for example, renal disease). What guidelines can be drawn from previous studies both of PAF-induced effects and of animal disease models that can help in designing new studies?

The water solubility and lack of irritancy of WEB 2086 and WEB 2170 make them suitable for intravenous or aerosol application, while the potency of these compounds not only when given by these routes but also when given orally or subcutaneously suggests good absorption characteristics. Receptor binding and *in vivo* data give some guideline to expected *in vivo* potencies (see Table 1), but as might be expected, the correlation between *in vitro* and *in vivo* is not perfect. In comparative studies with the other nonhetrazepine PAF antagonists

Table 2
Important Pharmacological Properties of WEB 2086, WEB 2170, and WEB 2347

	WEB 2086	WEB 2170	WEB 2347
Duration of action			
PAF-induced bronchoconstriction (guinea pig, po)	5.5 hours	12.1 hours	41 hours
PAF-induced hypotension (rat, po)	3.1 hours	5.4 hours	10.4 hours
PAF-induced hypotension (rat, IV)	About 1 hour	About 1 to 2 hours	3.5 hours
ED_{50}			
PAF-induced bronchoconstriction (guinea pig, po)	0.07 mg/kg	0.02 mg/kg	0.01 mg/kg
PAF-induced hypotension (guinea pig, po)	0.07 mg/kg	0.01 mg/kg	0.01 mg/kg
PAF-induced hypotension (rat, po)	9.2 mg/kg	0.26 mg/kg	0.1 mg/kg
PAF-induced hypotension (rat, IV)	0.06 mg/kg	0.02 mg/kg	0.003 mg/kg

Duration of action was calculated by giving a single dose of the PAF antagonist and measuring bronchoconstriction or hypotension after different periods. ED_{50} was calculated using the same conditions as in Table 1 (References 8, 56, 133, and Boehringer Ingelheim, internal data).

so far described, the hetrazepine PAF antagonists have consistently shown themselves to be one of the most potent group of substances presently available.[40 53]

Table 2 summarizes typical data on biological half-lives and potency for the most commonly used hetrazepine PAF antagonists.

Some points relating to this table are:

1. The biological half-life of the hetrazepine PAF antagonists in rats is considerably shorter than the biological half-life in guinea pigs. Preliminary data suggest that, in this respect, the situation in man is more "guinea pig-like" than "rat-like."

2. In comparison with the doses effective by intravenous administration, the effective doses for oral activity are much higher in rats than in guinea pigs.

3. WEB 2170, and to an even greater extent WEB 2347, are more potent than WEB 2086; possibly, a slower rate of metabolism might partly explain this greater potency.

4. Two of the most commonly measured biological effects of PAF *in vivo* are hypotension and bronchoconstriction. When both effects were measured (in response to an IV PAF challenge), the doses of WEB 2086 or WEB 2170 required to inhibit either effect were similar.

On the basis of these remarks, the following recommendations can be made for designing new tests of the effectiveness of WEB 2086 or WEB 2170 in disease models:

a. 24-hour inhibition of PAF effects is likely to exist when the following oral treatment regimens are used (all doses expressed per kilogram of body weight):

Rats

WEB 2086 b.i.d.	20mg am, 60mg pm
WEB 2170 b.i.d.	3mg am, 10mg pm
WEB 2347 b.i.d.	1mg am, 3mg pm

Guinea Pigs

WEB 2086 b.i.d.	3mg am, 10mg pm
WEB 2170 b.i.d.	1mg am, 3mg pm
WEB 2347	0.3mg daily

The doses are considerably higher than likely ED_{50} doses, but because the toxicity of these PAF antagonists is very low, such high doses can be used without a likelihood of side effects

b. Intravenous or aerosol administration are also suitable when lower doses are applicable.

c. WEB 2347, because of its higher partition coefficient, may be more suitable for studies in which ability to pass lipophilic barriers (for example, the blood-brain barrier) is required; despite its greater lipophilicity, specificity for the PAF receptor over the benzodiazepine receptor is such that no problems related to benzodiazepine-like effects on the CNS are likely.

The Clinical Perspective

The usefulness of PAF antagonists in disease can only be judged by properly controlled clinical trials. At the time of this writing, WEB 2086 has completed phase I trials (to test safety, tolerability, and PAF antagonistic activity in normal human volunteers) and is now in phase II trials (to test effectiveness against diseases in patients). These phase II trials are not yet complete.

In the phase I trials, which were all double-blind and placebo-controlled, WEB 2086 showed itself to be a potent and well-tolerated PAF antagonist whether given by the oral, intravenous, or inhaled route. In each study, *ex vivo* PAF-induced platelet aggregation (in most studies, 5×10^{-8}M PAF) has been used to monitor blocking of PAF receptors. ADP-induced aggregation has been measured at the same time as a control for nonspecific effects on platelet function.

Studies were:

1. A study with a single oral increasing dose of 1.25, 5, 20, 100, 200, and 400mg WEB 2086: marked (62 percent) inhibition of PAF-induced platelet aggregation was observed at the lowest dose tested, and from the dose 20mg upward, PAF-induced aggregation was totally inhibited, yet even the highest dose was well tolerated.[128]
2. A study with a single intravenous increasing dose (infusion of drug over 30 minutes at doses between 0.5mg and 50mg): in the lowest dose group (0.5mg), PAF-induced platelet aggregation *ex vivo* was attenuated by 68 percent, and doses of 10, 20, or 50mg provided complete inhibition.[129]

3. An inhaled drug study with a single increasing dose of 0.05, 0.25, 0.5, or 1mg WEB 2086.[129]

4. A time course and dose dependency study in which a single dose of 5, 30, or 90mg was administered orally, and blood samples were repeatedly obtained for measurement of PAF-induced platelet aggregation over a period of 24 hours. Maximal inhibition occurred between one and two hours. Magnitude and duration of inhibition was dose-dependent; a significant inhibition was still evident at ten hours after administration at all three dose levels and at 12 hours after administration at the two highest doses. Blockade of PAF-induced aggregation after an oral dose of 5mg/kg was reduced from maximal to 50 percent in approximately four to six hours. [130]

5. Two oral multiple dose studies (3 x 10mg/day or 3 x 40mg/day, dose interval eight hours) in which almost complete inhibition of PAF-induced aggregation occurred throughout the seven-day study period.[131]

In none of these studies were any clinically significant drug dose-related effects on either vital parameters (blood pressure, pulse rate, ECG, and so forth) or laboratory parameters (hematology, urinalysis, and serum chemistry) observed. All subjects completed the studies. In none of the studies did any of the subjects report any diazepine-characteristic side effects, even when a dose as high as 400mg was given or treatment was continued over seven days.

In addition to these studies showing *ex vivo* inhibition of PAF activity by WEB 2086, oral WEB 2086 (40mg single dose 1.5 hours prior to PAF challenge) was found to entirely block the bronchoconstrictive effects (change in airways resistance as measured by whole body plethysmography) of a PAF aerosol given to normal human volunteers.[48] This indicated that aerosolized PAF in humans as well as animals can exert its potent effects on the airways by an entirely receptor-mediated mechanism. The total prevention of PAF-induced bronchoconstriction by WEB 2086 contrasts with the only partial protection provided by the ginkgolide mixture BN 52063, even though the BN 52063 was dosed at levels able to significantly reduced PAF-induced wheal and flare reactions in the skin.[132]

The preclinical studies previously described support the clinical testing of hetrazepine PAF antagonists in treating a wide variety of diseases; these include diseases with an allergic or inflammatory component such as asthma or ulcerative colitis; diseases involving altered platelet function such as idiopathic thrombocytopenic purpura, and diseases in which changes in blood vessel diameter or permeability are important, for example shock and cardiac and cerebral ischemia. The results of such trials will be awaited with great interest and could lead to a novel approach in disease therapy.

References

1. J. Casals-Stenzel, *Naunyn-Schmiedebergs's Arch. Pharmacol.* **335**, 351 (1987).
2. J. Casals-Stenzel, K.H. Weber, G. Walther, and A. Harreus, FRG Patents DE 34 35 972 A1, DE 34 35 973 A1, DE 34 35 974 A1, 1984.
3. E. Kornecki, Y.H. Ehrlich, and R.H. Lenox, *Science* **226**, 1454 (1984).
4. W.E. Haefely, *Eur. Arch. Psychiatr. Neurol. Sci.* **238**, 294 (1989).
5. J.K.T. Wang, T. Taniguchi, and S. Spector, *Life Sci.* **27**, 1881 (1989).
6. T. Taniguchi, J.K.T. Wang, and S. Spector, *Life Sci.* **27**, 171 (1980).
7. J. Casals-Stenzel and K.H. Weber, *Br. J. Pharmacol.* **90**, 139 (1987).
8. J. Casals-Stenzel, G. Muacevic, and K.H. Weber, *J. Pharmacol. Exp. Ther.* **241**, 974 (1987).
9. K.H. Weber, G. Walther, A. Harreus, J. Casals-Stenzel, G. Muacevic, and W. Tröger, FRG Patent DE-OS 35 02 392 A1, 1986.
10. W. Stransky, K.H. Weber, G. Walther, A. Harreus, J. Casals-Stenzel, G. Muacevic, and W.D. Bechtel, FRG Patent DE 37 01 344 A1, 1987.
11. R.L. Rauchle *et al., Clin. Exp. Pharmacol. Physiol.* (Suppl. 14), Abstr. 16, (1989).
12. K.H. Weber and H.O. Heuer, *Int. Arch. Allergy Appl. Immunol.* **88**, 82 (1989).
13. K.H. Weber and H.O. Heuer, *Med. Res. Rev.* **9**, 181 (1989).
14. T. Tahara, M. Morikawa, M. Abe, and S. Yuasa, European Patent EP 268 242 A1, 1988.
15. A. Walser, European Patent EP 0 320 992 A2, 1989.
16. T.L. Yue *et al., Pharmacologist* **31**, 121 (1989).
17. P. Braquet and J.J. Godfroid, in *Platelet-Activating Factor and Related Lipid Mediators*, F. Snyder, Ed. (Plenum Press, New York, 1987), pp. 191-235.
18. D. Ukena *et al., FEBS Lett.* **228**, 285 (1988).
19. S.R. O'Donnell and C.J.K. Barnett, *Br. J. Pharmacol.* **94**, 437 (1988).
20. F.W. Birke *et al., Prostaglandins* **35**, 839 (1989).
21. L. Tahraoui *et al., Mol. Pharmacol.* **34**, 145 (1988).
22. R. Korth *et al., Br. J. Pharmacol.* **98**, 653 (1989).

23. K. Griffin, T. Hong, and J.V. Levy, *Biochem. Biophys. Res. Commun.* **160**, 263 (1989).
24. M. Schattner *et al., Br. J. Pharmacol.* **96**, 759 (1989).
25. H. Heuer *et al., Prostaglandins* **35**, 847 (1988).
26. K.J. Griffin and J.V. Levy, *Thromb. Res.* **51**, 219 (1988).
27. G. Dent *et al., FEBS Lett.* **244**, 365 (1989).
28. D. Ukena *et al., Biochem. Pharmacol.* **38**, 1702 (1989).
29. T. Fukuda *et al., J. Allergy Clin. Immunol.* **83**, 193 (1989).
30. C. Kroegel *et al., Immunology* **64**, 559 (1988).
31. C. Kroegel *et al., J. Immunol.* **142**, 3518 (1989).
32. C. Kroegel *et al., Biochem. Biophys. Res. Commun.* **162**, 511 (1989).
33. C. Kroegel *et al., FEBS Lett.* **243**, 41 (1989).
34. A.G. Stewart and G.J. Dusting, *Br. J. Pharmacol.* **94**, 1225 (1988).
35. S.G. Ward and J. Westwick, *Br. J. Pharmacol.* **93**, 769 (1988).
36. C. Kroegel, E.R. Chilvers, and P.J. Barnes, *Clin. Sci.* **76**, 45P (1989).
37. S-B. Hwang, M-H. Lam, and A.H-M. Hsu, *Mol. Pharmacol.* **35**, 48 (1989).
38. A.G. Stewart *et al., Br. J. Pharmacol.* **96**, 503 (1989).
39. A.G. Stewart and W.A. Phillips, *Br. J. Pharmacol.* **98**, 141 (1989).
40. P.G. Hellewell and T.J. Williams, *Br. J. Pharmacol.* **97**, 171 (1989).
41. A.M. Northover, *Agents Actions* **28**, 142 (1989).
42. M. Miwa, *et al., J. Clin. Invest.* **82**, 1983 (1988).
43. J.T. O'Flaherty *et al., J. Clin. Invest.* **78**, 381 (1986).
44. F.H. Valone, in *Platelet-Activating Factor and Related Lipid Mediators,* F. Snyder, Ed. (Plenum Press, New York, 1987) pp. 137-151.
45. G. Dent *et al., Am. Rev. Resp. Dis.* **139** (Suppl.), A94 (1989).
46. J. Gomez *et al., FASEB J.* **2**, 1577 (1988).
47. P.R.A. Johnson, C.L. Armour, and J.L. Black, *Eur. Resp. J.* **3**, 55 (1990).
48. W.S. Adamus *et al., Clin. Exp. Ther.* **47**, 456 (199).
49. H. Heuer, M. Schierenberg, and H. Darius, *Prostaglandins* **35**, 837 (1988).
50. B.B. Vargaftig *et al., Eur. J. Pharmacol.* **65**, 185 (1980).
51. J. Lefort, D. Rotilio, and B.B. Vargaftig, *Br. J. Pharmacol.* **82**, 565 (1984).
52. M. Pretolani *et al., Eur. J. Pharmacol.* **140**, 311 (1987).
53. H. Heuer *et al., Prostaglandins* **35**, 838 (1988).
54. H. Heuer *et al., Clin. Exp. Pharmacol. Physiol.* (Suppl. 13), Abstr. 7 (1988).
55. H. Heuer *et al., Prostaglandins* **35**, 798 (1988).
56. H.O. Heuer and K.H. Weber, *Lipids,* in press.
57. A. Lawson and I. Cavero, *Br. J. Pharmacol.* **93**, 76P (1988).
58. G. Anderson and M. Fennessy, *Br. J. Pharmacol.* **94**, 1115 (1988).
59. M. Del Monte and M. Subissi, *Naunyn Schmiedeberg's Arch. Pharmacol.* **338**, 417 (1988).
60. M. Pretolani, J. Lefort, and B.B. Vargaftig, *Br. J. Pharmacol.* **97**, 433 (1989).
61. J. Casals-Stenzel *et al., Br. J. Pharmacol.* **91**, 799 (1987).
62. L.J. Christy, A.G. Stewart, and G.J. Dusting, *Proc. Aust. Physiol. Pharmacol. Soc.* **19**, 94 (1988).
63. R.G. Goldie *et al., Clin. Exp. Pharmacol. Physiol.* (Suppl. 14), Abstr. 25 (1989).
64. S.R. O'Donnell, I.A.L. Erjefalt, and C.G.A. Persson, *Clin. Exp. Pharmacol. Physiol.* (Suppl. 14), Abstr. 70 (1989).

65. G. Muacevic, *Prostaglandins* **35**, 839 (1988).
66. T.W. Evans *et al., Br. J. Pharmacol.* **94**, 164 (1988).
67. S.R. O'Donnell and C.J.K. Barnett, *Clin. Exp. Pharmacol. Physiol.* (Suppl. 13), Abstr. 11 (1988).
68. A. Lellouch-Tubiana *et al., Am. Rev. Resp. Dis.* **137**, 948 (1988).
69. J. Homolka *et al., Am. Rev. Resp. Dis.* **135** (Suppl., part II), 160A (1987).
70. R.C. Read *et al., Am. Rev. Resp. Dis.* **139** (Suppl.), A481 (1989).
71. W.M. Abraham, J.S. Stevenson, and R. Garrido, *J. Appl. Physiol.* **66**, 2351 (1989).
72. H.O. Heuer, in *Bochumer Treff*, W.T. Ulmer, Ed. (Gedon and Reuss, Munich, in press).
73. E.J.A. Dixon *et al., Br. J. Pharmacol.* **97**, 717 (1989).
74. A. Fernandez *et al., Am. Rev. Resp. Dis.* **139** (Suppl.) A370 (1989).
75. I.C. Lohmann and M.Halonen, *FASEB J.* **2**,1577 (1988).
76. J. Casals-Stenzel, *Immunopharmacology* **13**, 117 (1987).
77. H.O. Heuer, *J. Allergy. Clin. Immunol.*, in press.
78. H. Heuer and J. Casals-Stenzel, *Agents Actions* (Suppl. 23), 207 (1988).
79. H. Darius *et al., Science* **232**, 58 (1986).
80. H. Darius, J.B. Smith, and A.M. Lefer, *Int. Arch. Allergy Appl. Immunol.* **80**, 369 (1986).
81. L.J. Christy, A.G. Stewart, and G.J. Dusting, *Proc. Aust. Physiol. Pharmacol. Soc.* **19**, 94 (1988).
82. J. Kleine-Tebbe, I. Schaefer, and G. Kunkel, *Allergology* **10**, 426 (1987).
83. A.M. Dunn, J.D. Palmer, and M. Halonen, *FASEB J.* **2**, 1577 (1988).
84. P.A. Hutson, S.T. Holgate, and M.K. Church, *Br. J. Pharmacol.* **95**, 770P (1988).
85. M. Soler, M. Sielczak, and W.M. Abraham, *J. Appl. Physiol.* **67**, 406 (1989).
86. D.F. Rogers, S. Dijk, and P.J. Barnes, *Am. Rev. Resp. Dis.* **137**, 236 (1988).
87. S-W. Chang, S. Fernyak, and N.F. Voelkel, *Am. J. Physiol.* **258**, H153 (1990).
88. A.W. Purvis *et al., Am. Rev. Resp. Dis.* **137** (Suppl.), 99 (1988).
89. J. Casals-Stenzel, *Eur. J. Pharmacol.* **135**, 117 (1987).
90. H.O. Heuer, *Naunyn Schmiedeberg's Arch. Pharmacol.* **339** (Suppl.), R78 (1989).
91. H.O. Heuer, in *Immune Responses and Renal Diseases*, P. Braquet *et al.*, Eds. (Excerpta Medica, Hong Kong, 1989), pp. 48-56.
92. D.G. McCormack, P.J. Barnes, and T.W. Evans, *Science* **77**, 439 (1989).
93. D.G. McCormack, P.J. Barnes, and T.W. Evans, *Clin. Sci.* **76**, 44P (1989).
94. C. Chen, N.F. Voelkel, and S. Chang, *Am. Rev. Resp. Dis.* **139** (Suppl.), A273 (1989).
95. J.V. Esplugues and B.J.R. Whittle, *Meth. Find. Exp. Clin. Pharmacol.* **11** (Suppl. 1), 61 (1989).
96. A. Brambilla, A. Ghiorzi, and A. Giachetti, *Pharmacol. Res. Commun.* **19**, 147 (1987).
97. H. Heuer, J. Casals-Stenzel, and K.H. Weber, *Clin. Exp. Pharmacol. Physiol.* (Suppl. 13), Abstr. 50 (1988).
98. H.O. Heuer, G. Letts, and C.J. Meade, *J. Lipid Mediators* **2**, S101 (1990).
99. J.N. King and E.L. Gerring, *Br. J. Pharmacol.* **97**, 428P (1989).
100. T.L. Buckley and J.R. Hoult, *Eur. J. Pharmacol.* **163**, 275 (1989).
101. J.L. Wallace, G.C. Ibbotson, and C.M. Keenan, in *Ginkgolides–Chemistry, Biology, Pharmacology, and Clinical Perspectives, Vol. 2*, P. Braquet, Ed. (1989, Prous, Barcelona, in press).
102. M. Sanchez-Crespo *et al., Immunopharmacology* **4**, 173 (1982).
103. J. English and P.D. Toth, *Prostaglandins* **35**, 825 (1988).

104. A.W.B. Stanton *et al., Br. J. Pharmacol.* **97**, 643 (1989).
105. V. Lagente *et al., J. Pharmacol. Exp. Ther.* **247**, 254 (1988).
106. V. Lagente *et al., Clin. Exp. Pharmacol. Physiol.* (Suppl. 13), Abstr. 32 (1987).
107. H.W. Meining, A.G. Stewart, and G.J. Dusting, *Clin. Exp. Pharmacol. Physiol.* (Suppl. 12), Abstr. 22 (1988).
108. A.G. Stewart and G.J. Dusting, *Clin. Exp. Pharmacol. Physiol.* (Suppl. 13), Abstr. 31 (1988).
109. S.B. Felix *et al., Z. Kardiol.* **78** (Suppl.), Abstr. 36 (1989).
110. H.B. Yaacob and P.J. Piper, *Br. J. Pharmacol.* **95**, 521P (1988).
111. S.B. Felix *et al., Eur. Heart J.* **9** (Suppl.), Abstr. 171 (1988).
112. H.J. Mest *et al., Biomed. Biochim. Acta* **47**, 219 (1988).
113. E. Schroeder *et al., Circulation* **78** (Suppl.), 77 (1988).
114. H. Pouleur, E. Schroeder, and H. Van Mechelen, Effects of dipyridamole and of the platelet-activating factor antagonist WEB-2086 on post-reperfusion myocardial damage in a low flow, high demand model of ischemia (privately communicated document, 1990).
115. P.D. Collins *et al., Br. J. Pharmacol.* **94**, 409P (1988).
116. M.A. Martins *et al., Eur. J. Pharmacol.* **149**, 89 (1988).
117. H. Heuer *et al., Lipids*, in press.
118. F. Bussolino *et al., Thromb. Res.* **48**, 619 (1987).
119. H.O. Heuer, *Naunyn Schmiedeberg's Arch. Pharmacol.* **337** (Suppl.), R70 (1988).
120. H. Siebeck *et al., Am. Rev. Resp. Dis.*, in press.
121. G.L. Stahl, H. Bitterman, and A.M. Lefer AM, *Thromb. Res.* **53**, 327 (1989).
122. G.W. Bielenberg and G. Wagener, in *Pharmacology of Cerebral Ischemia*, J. Krieglstein, Ed, (CRC Press, Boca Raton, Florida, 1988), pp. 281-284.
123. B. Spinnewyn, *Prostaglandins* **34**, 337 (1987).
124. S.H. Peers, *Eur. J. Pharmacol.* **150**, 131 (1988).
125. M.A. Martins *et al., Br. J. Pharmacol.* **96**, 363 (1989).
126. R.S.B. Cordeiro *et al., Prostaglandins* **35**, 836 (1988).
127. M. Da Costa, S. Metcalfe, and R.Y. Calne, *Transplant. Proc.*, in press.
128. W.S. Adamus *et al., Prostaglandins* **35**, 797 (1988).
129. W.S. Adamus, H. Heuer, and C.J. Meade, *Meth. Find. Exp. Clin. Pharmacol.* **11**, 415 (1989).
130. W.S. Adamus *et al., Eur. J. Clin. Pharmacol.* **35**, 237 (1988).
131. W.S. Adamus *et al., Clin. Pharmacol. Ther.* **45**, 270 (1989).
132. N.M. Roberts *et al., Br. J. Clin. Pharmacol.* **26**, 65 (1988).
133. H.O. Heuer and K.H. Weber, *Naunyn Schmiedeberg's Arch. Pharmacol.* **340** (Suppl.), R75 (1989).

The Pathophysiology of PAF

5

Platelet-Activating Factor as a Mediator in Pathophysiology of the Central Nervous System

Tian-Li Yue
Department of Pharmacology
SmithKline Beecham

Kai U. Frerichs
Perttu J. Lindsberg
Department of Neurology
Research Division
Uniformed Services
University of the
Health Sciences
Bethesda, MD

Pierre Braquet
Institut Henri Beaufour
Paris, France

Paul G. Lysko
Richard M. Edwards
R. Rabinovici
correspondence
Giora Feuerstein
Department of Pharmacology
SmithKline Beecham
P.O. Box 1539, L-511
King of Prussia, PA 19406

Platelet-activating factor (PAF) is known to be produced and act on a variety of cell and tissue types.[1,2] The brain is an organ rich in phospholipids and contains relatively high levels of enzymes for the synthesis and metabolism of alkylether phospholipids.[3] More than 4 percent of brain phospholipids are ether lipids, half of which represent glycerolether phospholipids with a phosphatidylcholine moiety, a potential PAF precursor.[4,5] Therefore, it is of interest to explore the possibility of such a lipid mediator being produced in the central nervous system (CNS) and its functions under normal and abnormal conditions. In this brief review, we present the current knowledge in support of PAF presence in the CNS and its potential role in CNS pathological situations particularly in stroke and neurotrauma.

PAF Production in the Central Nervous System

It was first reported in 1976 that a phospholipid isolated from the lipid fraction of the bovine brain caused hypotension; this active principle was termed "Depressor-I."[6-9] One purified subfraction strongly aggregated platelets and caused mild hypotension; both effects were blocked by the platelet-activating factor (PAF) antagonist CV-3988.[10] This fraction was further analyzed by gas chromatography and mass spectrometry (GC-MS) and found to be composed of four species of PAF (16:0, 16:1, 18:0, and 18:1 PAF) and ten acyl analogues of PAF. The level of 16:0 and 18:0 PAF was 2.5μg and 7.5μg per 400 to 450g cerebrum, respectively. The total amount (μg/cerebrum) of acyl analogues of PAF was 55 times higher than that of PAF.[11] Recently, the same group,[6-9] using high-resolution capillary GLC-MS, found an additional seven compounds of acyl PAF in bovine brain, which have not previously been detected in natural samples.[12]

In 1988, Kumar *et al.*[13] reported that a detectable PAF activity was observed in the rat brain using bioassay techniques. The low basal level of PAF (0.25 ± 0.15pmol/g) in the brain was greatly increased by chemoconvulsion or electroconvulsion.[13] Similar results were found in electroconvulsive mice, in which the existence of PAF in brain was confirmed by GC-MS, and the amount of PAF in the brain quickly decreased as the animal recovered subsequent to stimulation.[14] In our preliminary studies of rat brain, PAF was identified following extraction from tissue with C_{18} cartridge (Bond Elut, Analytichem) and purified further with LC-Si cartridge (Supelclean, Supelco™) and thin-layer chromatography (TLC) and finally measured by GC-MS. The level of PAF in cortex was 3.67 ± 0.20ng/g, which was similar to that reported in bovine brain.[11] It is interesting that the level of PAF in brain stem was almost twofold greater than that present in cortex. Using bioassay technique ([^{3}H]serotonin release),[15] we also observed the activity of PAF in rabbit spinal cord after partial purification of the sample by TLC. The basal level of PAF in the spinal cord was significantly elevated with ischemia-induced damage.[16]

It is of interest to investigate whether PAF is produced within or transported into the brain parenchyma. Bussolino *et al.*[17] demonstrated that PAF could be produced by chick retina, a tissue structure used for the investigation of cells of CNS, upon stimulation with neurotransmitters such as acetylcholine, dopamine, or calcium ionophore A23187. In isolated perfused rat brain exposed to chemoconvulsants, Kumar *et al.*[13] indicated that production of PAF in the brain is independent of systemic metabolism. Recently, using cultured rat cerebellar granule cells as a model, we observed that PAF could be produced by these cells with or without appropriate stimuli.[19] In these latter studies, PAF was measured by serotonin release test after partial purification of samples by TLC, and this bioactivity due to PAF was confirmed by the following evidence:

(1) the bioactivities, both serotonin release as well as platelet aggregation, were inhibited by specific PAF antagonist BN 50739;[18]

(2) base-catalyzed methanolysis or pretreatment with phospholipase C completely abrogated this bioactivity;

(3) lysoPAF present in the sample transformed into PAF by chemical acetylation reconstructed the bioactivity;

(4) finally, normal-phase high-performance liquid chromatography of a pooled sample showed that the main peak with the highest bioactivity was eluted at the same retention time as the standard [^{3}H]-PAF (Figure 1). PAF was detected both in the cells and the supernatant at approximately equal levels, indicating that PAF was indeed released from the cerebellar granule cells.[19]

PAF can be enzymatically synthesized by either a remodeling or a *de novo* pathway.[2] Because acetyltransferase activity was not detected in rat brain,[13,20] it was suggested that the remodeling pathway seemed to be inoperative in brain tissue because of the lack of acetyltransferase. The last step of the *de novo* pathway in the rat brain has been confirmed recently.[21] These researchers[21] demonstrated that rat brain is able to synthesize PAF from 1-alkyl-2-acetyl-*sn*-glycerol

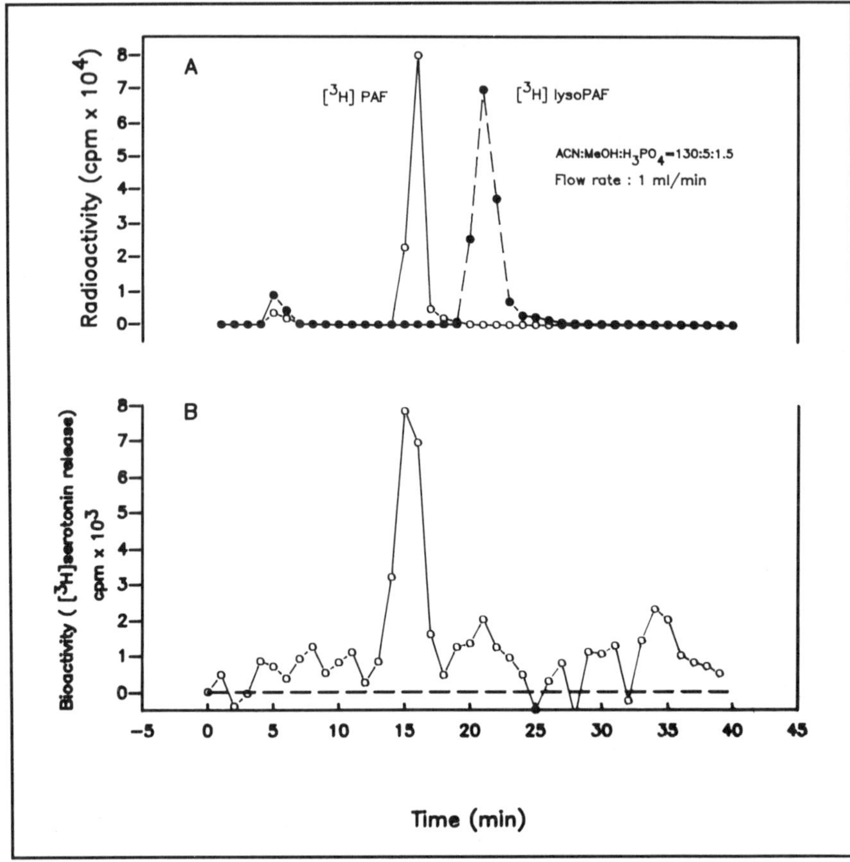

Figure 1. High performance liquid chromatographic (HPLC) identification of PAF from cultured rat cerebellar granule cells. **A:** Separation of standard [^{3}H]-PAF and [^{3} H]-lysoPAF by HPLC; **B:** Measurement of bioactivity (serotonin release test). Granule cell culture was extracted and purified by thin-layer chromatography, followed by HPLC. Each fraction (one fraction per minute) was collected, and bioactivity was measured.

and cytidine 5'-diphosphate (CDP)-choline by a "dithiothreitol (DTT)-insensitive" phosphocholine transferase, which transfers the phosphocholine residue from CDP-choline to the lipid acceptor, 1-alkyl-2-acetyl-*sn*-glycerol. The phosphocholine transferase enzyme has a microsomal localization, requires Mg^{++} and is inhibited by Ca^{2+}. These results provide evidence that nervous tissue possesses the capability of synthesizing this lipid mediator by the *de novo* pathway.

Specific Binding Sites for PAF in the Brain

The possibility that PAF acts in the CNS is supported by the presence of binding sites for both gerbil and rat brain.

Specific, reversible and saturable binding sites for [^{3}H]-PAF were observed in gerbil brain. Scatchard plot analysis revealed the existence of two apparent classes of binding sites with affinity coefficients $K_{d1} = 3.66 \pm 0.92$nM and $K_{d2} = 20.4 \pm 0.50$nM, corresponding respectively to a maximum number of binding sites: $Bmax_1 = 0.83 \pm 0.23$pmol/mg protein and $Bmax_2 = 1.1 \pm 0.32$pmol/mg protein. The binding of [^{3}H]-PAF was fully displaced by cold PAF and partially inhibited by PAF antagonist BN 52021, kadsurenone, and L-652731, suggesting that these antagonists might interact only with one binding site. Distribution of [^{3}H]-PAF specific binding revealed a maximum density of binding sites in midbrain and hippocampus.[22]

In similar studies,[23] two specific binding sites for [^{3}H]-PAF have been found and characterized in rat hypothalamus. The K_d of each binding site was 2.14 ± 0.32 and 61.63 ± 16.40nM, respectively, and the corresponding maximal number of each binding site was 25.4 ± 3.2 and 146.2 ± 47.5fmol/mg protein. In the same conditions, no specific binding was observed using rat pituitary membranes. The specificity of PAF analogues for these binding sites was well correlated to their relative effectiveness in altering luteinizing hormone releasing hormone (LHRH) and somatostatin (SRIF) release, suggesting that PAF effect on neuropeptides release may be a receptor-mediated process.[23,24] Recently, specific binding sites for [^{3}H]-PAF in rat brain cortex were also found, and the highest specific binding sites were in the microsomal fraction.[25]

Effects of PAF in the CNS

Direct infusion of PAF into the rat carotid artery (67pmol/minute, for 60 minutes) was shown to reduce cerebral blood flow (CBF) by 25 percent and to decrease systemic blood pressure by 37 percent. In contrast, the global cerebral metabolic rate for oxygen

($CMRO_2$) increased after 15 minutes. There was no significant change in the CBF or $CMRO_2$ in control animals in which systemic hypotension was induced by bleeding to verify that the CBF change was unrelated to systemic hemodynamics.[26] The results were strikingly similar to those observed in endotoxin-treated animals[27] and during the delayed phase of postischemic reperfusion.[28,29] Moreover, Kochanek *et al.*[30] reported recently that PAF antagonists BN 52021 and WEB 2086 failed to alter normal CBF in rats, whereas indomethacin markedly reduced normal CBF, which suggests that CBF seemed not to be modulated by PAF under normal conditions. Consistent with this finding are our recent observations on the effect of PAF on isolated cerebral arterioles. Single intracerebral arterioles (9.3 to 49.5μm) were dissected from normal rats, and changes in lumen diameter were measured using techniques similar to those reported by Dacey and Duling.[31] PAF added to the abluminal side of the arterioles had no effect on lumen diameter (Figure 2). The arterioles were clearly intact because they relaxed in response to the vasodilator calcitonin gene related peptide and contracted in response to the thromboxane mimetic U44619. However, a direct vasoconstrictor effect of PAF on pial arterioles was observed by Armstead *et al.*[32] Using a closed window technique in newborn piglets, PAF superfusion produced dose-dependent decrease in pial arteriolar diameter. Because there was no increase in the levels of prostaglandins, the effect seemed not to relate to the formation of arachidonic acid metabolites. The differences between pig pial vessels and the rat cortical microvessel response to PAF might be explained by species differences and/or the difference between extraparenchymal versus intraparenchymal vessels.

PAF injected directly into brain parenchyma increased blood-brain barrier permeability in a similar fashion to PAF effect on peripheral vessels.[33] Kumar *et al.*[13] found that PAF infused into the vasculature of the isolated perfused rat brain caused changes consistent with an increase in blood-brain barrier permeability, although PAF did not by itself significantly penetrate the blood-brain barrier. Moreover, these authors observed that when a ^{32}P-labeled synaptosome preparation from rat brain was challenged with PAF at 0.1nM

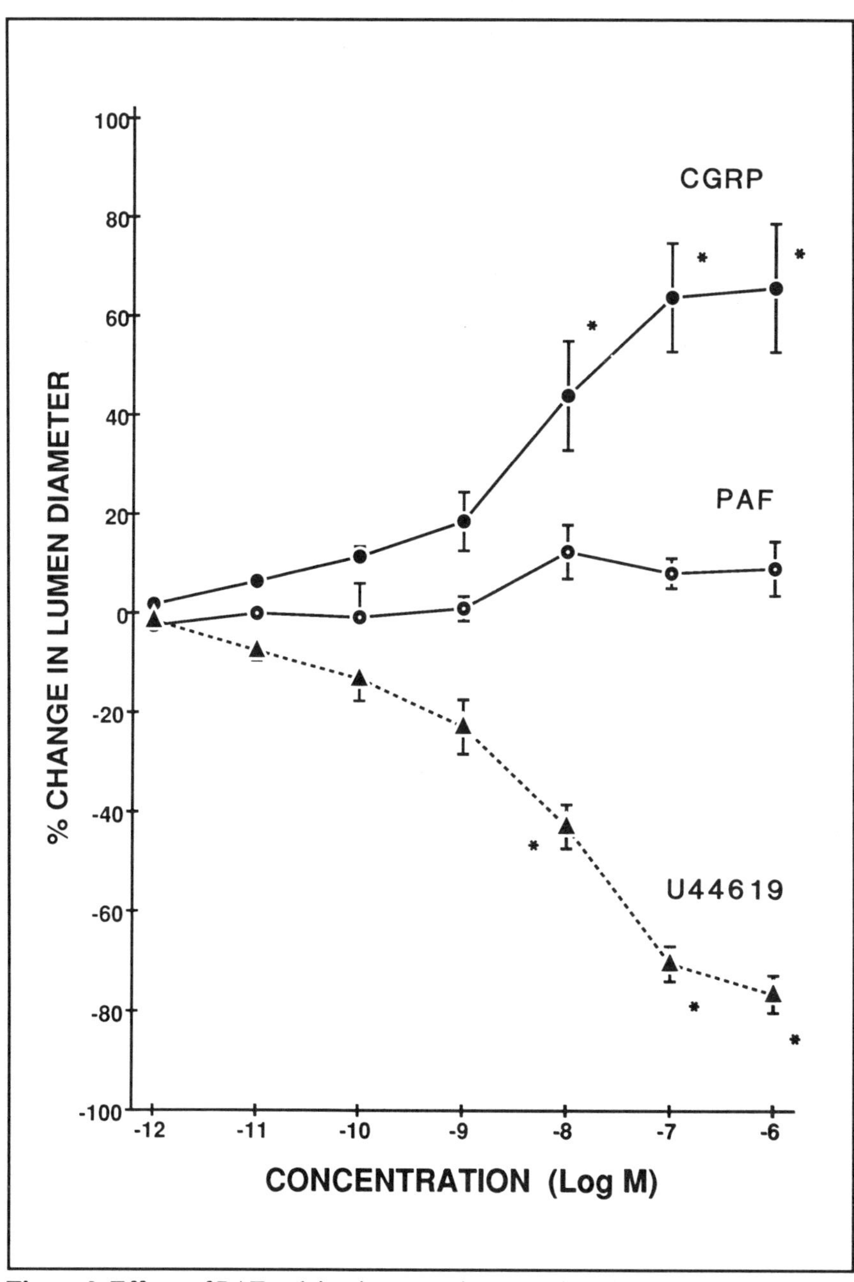

Figure 2. Effects of PAF, calcitonin gene related peptide (CGRP), and U44619 on lumen diameter of rat intracerebral arterioles. Results are expressed as percent change in central lumen diameter that was 20.6 ± 2.7μm, *n*=15. Asterisk denotes significant (*P*<0.05) change.

concentration, the turnover of polyphosphoinositides was accelerated and the Na^{+}-Ca^{2+} exchange of synaptic membrane vesicles was increased.

Using laser-Doppler flowmetry, the effect of PAF on rabbit spinal cord microcirculation (SCM) was observed recently in our laboratory. The spinal cord blood flow (SCBF) decreased by 14 ± 5 percent ($P<0.05$) during the infusion of PAF, paralleling the systemic hypotensive changes (17 ± 5 percent, $P<0.05$), with no changes in vascular resistance (SCVR). However, immediately after termination of PAF infusion, SCVR decreased, whereas SCBF rapidly recovered (Figure 3). Indomethacin blocked both the hemodynamic events and the eicosanoid release induced by PAF, suggesting that PAF modulates SCM through an eicosanoid-mediated mechanism.[34]

Neuroendocrine Effects of PAF

PAF decreased LHRH and SRIF release from the rat median eminence *in vitro,* with a maximal inhibition at 10^{-14}M for both neuropeptides, whereas the release of growth hormone releasing factor (GRF) was not significantly altered. Moreover, PAF strongly counteracted the Ca^{2+} ionophore A23187-stimulated release of LHRH and SRIF from median eminence and medial basal hypothalamus (>50 percent inhibition), and the effect of PAF on neuropeptides release was altered by specific PAF antagonists. Besides, this inhibitory effect was specifically exerted at a hypothalamic site, because PAF failed to depress luteinizing hormone (LH) and growth hormone (GH) release from anterior pituitary.[23] Camoratto and Grandison[35] reported that PAF stimulated prolactin (PRL) release from dispersed rat anterior pituitary cells *in vitro,* which was dose-dependent and blocked by PAF antagonists and dopamine agonists. In addition, PAF induced the secretion of PRL and GH but not that of LH or thyroid-stimulating hormone from hemipituitaries in short-term incubation.[35]

Kornecki and Ehrlich[36] found that exposure of NG 108-15 neuronal cells to low concentrations of PAF (50nM to 2.5μM) induced morphological differentiation with neurite extension; however, higher

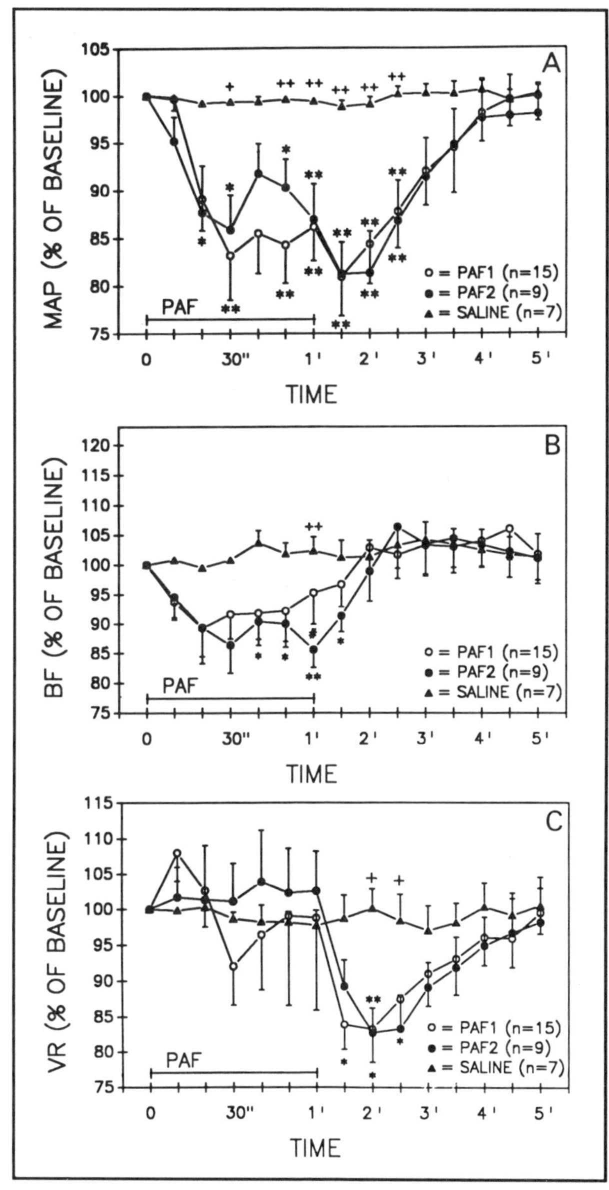

Figure 3. Response of hemodynamic parameters to two consecutive infusions of PAF separated by 50 minutes (IV infusion. 0.5nmol/kg). **A:** Responses of mean arterial pressure (MAP); **B:** Responses of spinal cord blood flow (SCBF); **C:** Responses of spinal cord vascular resistance (SCVR). $^{*}P <$ 0.05; $^{**}P < 0.01$, versus saline control. # in **B** indicates a significant difference between the two PAF-groups ($P < 0.05$).

concentrations of PAF (>3.5μM) were neurotoxic. PAF increased the intracellular levels of calcium ions in cells and this effect was dependent on extracellular calcium and inhibited by PAF antagonists. It was suggested from these observations that PAF may play a physiological role in neuronal development and a pathophysiological role in the degeneration that occurs in trauma, stroke, or spinal cord injury. The significance of calcium in the PAF-mediated neurotropic effects was also confirmed by a recent study in PC-12 cell line. The R-form steric configuration of PAF but not its S-form induced dopamine release, which was associated with a rise in free cytosolic calcium that was blocked by PAF antagonist.[37]

PAF as a Mediator in Pathological CNS States and the Therapeutic Capacity of Selected PAF Antagonists

So far, the potential involvement of PAF in the complex pathophysiology of CNS injury was mainly inferred from studies where the beneficial effects of the specific PAF antagonists were obtained, and only in a few cases the increased concentration of PAF was also observed. The latter observation is crucial to confirmation of the hypothesis that PAF is involved in certain pathological CNS states.

PAF as a Mediator of Pathophysiology of Ischemia and Reperfusion-Induced Neuronal Injury

A growing body of evidence implicates PAF as playing an important role in ischemia and reperfusion-induced neuronal damage. In the gerbil model of global cerebral ischemia (bilateral carotid clamping), it was found that the PAF antagonist BN 52021 was able to ameliorate the neurological deficits after ischemia in a dose-related manner as assessed by the stroke index;[38] this finding was paralleled by an improved mitochondrial respiration rate (Figure 4). Not only pretreatment but also curative administration one hour after ten minutes of cerebral ischemia resulted in improved trend of cerebral function. However, the initial cerebral impairment was not modified by BN 52021 administered preventively, suggesting that PAF antagonists improved the postischemia phase without effect on initial cerebral impairment. Panetta *et al.*[39] demonstrated in the same is-

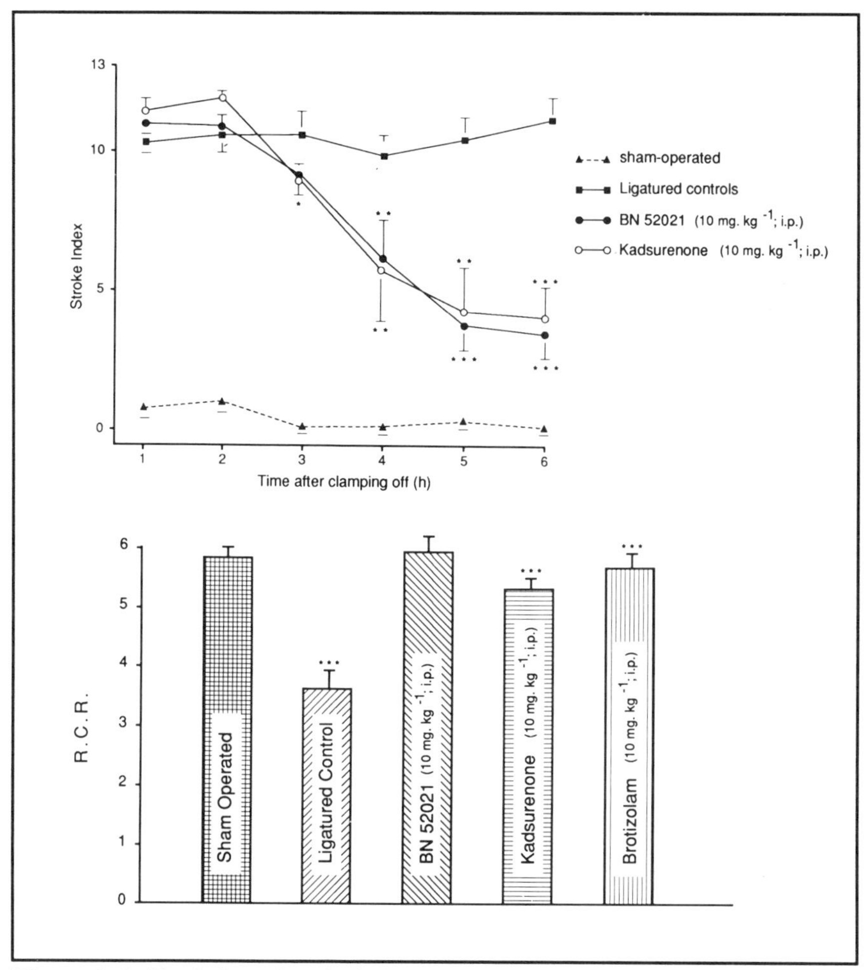

Figure 4. A: Evolution of stroke index recorded after ten minutes of ischemia and recirculation in conscious Mongolian gerbils; **B:** Respiratory control ratio (RCR) in gerbil brain mitochondria following ten minutes of carotid ligation and six hours recirculation. Bn 52021, kadsurenone, and brotizolam were given one hour after clamping off. * $P<0.05$; ** $P<0.01$; *** $P<0.001$ versus control ligatured gerbils.

chemia model inhibition of evolution of ischemic injury by BN 52021; animals treated with the PAF antagonist on reperfusion had an improved CBF after 60 and 90 minutes. At 90 minutes after ischemia, there was almost complete recovery of CBF in forebrain and midbrain, whereas the vehicle-treated control group showed the typical hypoperfusion pattern. In addition, BN 52021 treatment decreased the pool of free fatty acids (FFA) in the forebrain.[39] The latter effect of BN 52021 was also confirmed in mice. PAF antagonists, given 30 minutes before electroconvulsive shock or postdecapitation ischemia, significantly inhibited FFA accumulation in both injury models. These data suggest that PAF might be involved in the injury-induced activation of phospholipase A_2.[40]

In a model of multifocal cerebral ischemia by air embolism in dogs, the effect of pretreatment with the PAF antagonist kadsurenone on neuronal recovery, platelet accumulation, and CBF was assessed. The neuronal recovery (monitored by cortical somatosensory evoked potentials) was improved by pretreatment with kadsurenone, whereas no changes were found in CBF or platelet accumulation. The beneficial effect of the PAF inhibition might have been related to antagonism of leukocyte aggregation and activation by PAF, which could have altered the blood-brain barrier as well as the microcirculation.[41] Braquet *et al.*[25] reported recently that PAF at low doses had the capacity to aggravate ischemia-induced neuronal damage in mice and gerbils (ischemia-aggravating effect). Unfortunately, the level of PAF in the brain before and after ischemia and reperfusion was not studied in all these models.

Our laboratory modified the well-described model of rabbit lumbar spinal cord deteriorating stroke (SCS) to allow on-line monitoring of spinal cord blood flow (BF) with laser-Doppler flowmetry through a laminectomy at L5.[16] Using this model, the level of PAF in ischemic spinal cord tissue was observed. SCS was induced by inflating a pediatric balloon-tipped Swan-Ganz catheter advanced via the right femoral artery to occlude the abdominal aorta directly below the renal arteries. Mean arterial pressure (MAP) proximal to the occlusion was monitored. At the end of each experiment, spinal cord samples were excised promptly from the ischemic lumbar area and nonischemic upper thoracic sections for PAF assay and water content measurement. The timepoint for tissue collection was selected by two hours of reperfusion, which coincided with the development of delayed hypoperfusion. Bioassay of PAF based on the PAF-induced serotonin release from washed rabbit platelets after partial purification of sample by TLC and the bioactivity due to PAF were confirmed as described previously. PAF levels in ischemic sections of the spinal cord were substantially higher than the levels in nonischemic samples from the same animals. PAF antagonist BN 50739 effectively blocked the delayed hypoperfusion, whereas the initial reperfusion hyperemia induced by ischemia was not affected. Moreover, pretreatment with BN 50739 significantly decreased postischemic edema formation.[42] Our results support the hypothesis that PAF plays an important role in ischemia and reperfusion-induced neuronal injury.

In a recent clinical study, it was shown that the activity of PAF acetylhydrolase was higher in the plasma of patients with ischemic stroke than that in healthy controls. The increased enzyme level in the stroke patients was correlated with an elevated PAF-induced platelet aggregation. The study suggested that the platelet hyperfunction might be associated with an augmented generation of PAF, which in turn would induce the inactivating enzyme acetylhydrolase.[43]

Neurotrauma

Buchanan *et al.*[44] reported that BN 52021 attenuated the edema formation (hemispheric right-left difference in percent brain water) in rats subjected to a 50g x cm^2 traumatic insult, although there was no difference in any of the controlled parameters of intracranial pressure before or after trauma. They speculated that PAF may be involved in the posttraumatic cerebral edema formation. Faden *et al.*[45] reported recently that fluid-percussion-induced traumatic brain injury caused an approximate twofold increase in PAF level, as measured by the rabbit platelet aggregation test, in rat brain tissue.

Pretreatment with BN 52021 showed a dose-related improvement in behavioral recovery at two weeks post trauma.[45] Using the same method to induce brain injury, Vishwanath *et al.*[46] found that traumatic brain injury caused changes in phospholipid phosphorous associated with activation of acid and neutral phospholipase C.

In recent studies[47] in our laboratory using a neodymium:yttrium-aluminum-garnet (ND:YAG) laser to induce focal cortical lesion in anesthetized rats, the beneficial prophylactic effect of PAF antagonist BN 50739 on brain edema, cortical microcirculation, blood-brain barrier disruption, and neuronal death following focal brain injury was observed. The histopathological evolution was followed up to four days. Neuronal damage in the cortex and the hippocampus (CA-1) was assessed quantitatively, revealing secondary and progressive loss of neuronal tissue within the first 24 hours following injury. Pretreatment with the potent and selective PAF antagonist BN 50739 ameliorated the severe hypoperfusion and reduced edema acutely after injury. This PAF antagonist also reduced the progression of neuronal damage in the cortex and CA-1 hippocampal neurons. The results of these studies provide evidence in support of progressive brain damage following focal brain injury, which is associated with secondary loss of neuronal cells; PAF might be a key mediator of such secondary brain damage following cerebral neurotrauma.[47]

Convulsive Disorders

Like ischemic reperfusion injury, electroconvulsive shock results in loss of FFA and mainly stearic and arachidonic acids from phosphatidylinositol-4,5-bis-phosphate (PIP_2) in mouse brain. Pretreatment with BN 52021 reduced the accumulation of these FFA, with no effect on the content or composition of fatty acids in diacylglycerol, nor was there a loss of PIP_2, indicating that phospholipase C was not activated.[40] Kumar *et al.*[13] reported that intraperitoneal injection of the chemoconvulsant agents (picrotoxin or bicuculline) or electroconvulsion significantly increased PAF levels in rat brain from the basal level of 0.25 ± 0.15pmol/g to 10.68 ± 2.18, 4.97 ± 0.75, and 1.76 ± 0.30pmol/g, respectively. However, whether PAF antagonists have any preventive or therapeutic effects in convulsive disorders remains a question.

Conclusion

Despite the rapidly expanding interest and effort to investigate PAF and its role in CNS pathophysiological situations, the exact mechanisms of its physiological and pathological actions are still not well understood, and the evaluation of the molecular and biochemical mechanisms requires additional experimental efforts. Moreover, the potential therapeutical effects of PAF antagonists in different pathological states of CNS need further experimental elucidation, which will allow better understanding of the importance of PAF in CNS function and may provide new opportunities in the treatment of CNS disorders.

References

1. P. Braquet *et al., Pharmacol. Rev.* **39**, 97 (1987).
2. F. Snyder, *Proc. Soc. Exp. Biol. Med.* **190**, 125 (1989).
3. M. Blank *et al., J. Biol. Chem.* **256**, 175 (1981).
4. G. Clarke and C. Dawson, *Biochem. J.* **195**, 301 (1981).
5. L.A. Horrocks and M. Sharma, in *Phospholipids,* J.N. Howthorne and G.B. Ansell, Eds. (Elsevier Biomedical Press, New York, 1982), pp. 51-93.
6. H. Tsukatani *et al., Chem. Pharm. Bull.* **24**, 2294 (1976).
7. H. Tsukatani *et al., Chem. Pharm. Bull.* **26**, 3271 (1978a).
8. H. Tsukatani *et al., Chem. Pharm. Bull.* **25**, 3281 (1978b).
9. H. Tsukatani *et al., Jap. J. Pharmacol.* **29**, 695 (1979).
10. J.-I. Yoshida *et al., J. Pharm. Pharmacol.* **38**, 878 (1986).
11. A. Tokumura, K. Kamiyasu, and H. Tsukatani, *Biochem. Biophys. Res. Commun.* **145**, 415 (1987).
12. A. Tokamura *et al., J. Lipid Res.* **30**, 219 (1989).
13. R. Kumar *et al., Biochim. Biophys. Acta* **963**, 375 (1988).
14. K. Clay and R. Baker, in *Proceedings of the 33rd Annual Conference of Mass Spectrometry and Allied Topics,* (San Diego, CA, 1985), pp. 700-701.
15. R.N. Picckard, R.S. Farr, and D.J. Hanahan, *J. Immunol.* **123**, 1947 (1979).
16. P. Lindsberg *et al., Soc. Neurosci.* Abstr. **15**, 805 (1989).
17. F. Bussolin *et al., J. Biol. Chem.* **261**, 16502 (1986).
18. T.L. Yue *et al., Pharmacologist* **31**, 121 (1989).
19. T.L. Yue, P.G. Lysko, and G. Feuerstein, *Neurochemistry,* in press.
20. R.L. Wykle, B. Malone, and F. Snyder, *J. Biol. Chem.* **255**, 10256 (1980).
21. E. Francescangeli and G. Goracci, *Biochem. Biophys. Res. Commun.* **161**, 107 (1989).
22. M.T. Domingo *et al., Biochim. Biophys. Res. Commun.* **151**, 730 (1988).
23. M.P. Junier *et al., Endocrinology* **123**, 72 (1988).
24. C. Rougest *et al., Prostaglandins* **35,** 816 (1988).
25. P. Braquet *et al.*, in *Platelet-Activating Factor and Diseases,* K. Saito and D. Hanahan, Eds. (International Medical Publishers, Tokyo, Japan, 1989), pp. 85-102.
26. P.M. Kochanek *et al., J. Cereb. Blood Flow Metab.* **8**, 546 (1988).
27. B. Ekstrom-Jodal, E. Haggendal, and L.E. Larsson, *Acta. Anaesthesiol. Scand.* **26**, 163 (1982).
28. J.V. Snyder *et al., Stroke* **6**, 21 (1975).

29. E. Nemoto, K.-A. Hossmann, and H. Cooper, *Stroke* **12**, 66 (1981).
30. P. Kochanek *et al., Stroke* **20,** 134 (1989).
31. R.G. Dacey and B.R. Duling, *Am. J. Physiol.* **253**, H1253 (1987).
32. W. Armstead *et al., Circ. Res.* **62**, 1 (1988).
33. M. Plotkine *et al.*, in *Ginkgolides-Chemistry, Biology, Pharmacology and Clinical Perspective*, P. Braquet, Ed. (Prous, Barcelona, Spain, 1988), pp. 687-689.
34. P.J. Lindsberg *et al., J. Lip. Med.* **2**, 41 (1990).
35. A.M. Camoratto and L. Grandison, *Endocrinology* **124,** 1502 (1989).
36. E. Kornecki and Y.H. Ehrlich, *Science* **240,** 1792 (1988).
37. F. Bussolin, F. Tessari, and F. Turrini, *Am. J. Physiol.* **255**, C559 (1988).
38. B. Spinnewyn *et al., Prostaglandins* **34**, 337 (1987).
39. T. Panetta *et al., Biochem. Biophys. Res. Commun.* **149**, 580 (1987).
40. D.L. Birkle *et al., Prostaglandins* **35**, 831 (1988).
41. P.M. Kochanek *et al., Life Sci.* **41**, 2639 (1987).
42. P.J. Lindsberg *et al., Stroke,* in press.
43. K. Satoh *et al., Prostaglandins* **35**, 688 (1988).
44. D.C. Buchanan *et al., Prostaglandins* **35**, 814 (1988).
45. A.I. Faden *et al., Soc. Neurosci.* Abstr. **15**, 1112 (1989).
46. B. Vishwanath *et al., Soc. Neurosci.* Abstr. **15**, 133 (1989).
47. K.U. Frerichs *et al., J. Neurosurg*., in press.

6

PAF Antagonists and the Immune Response

Jean Michel Mencia-Huerta
Pierre Braquet
Institut Henri Beaufour
1, avenue des Tropiques
91952 Les Ulis
France

correspondence
Benjamin Bonavida
Department of Microbiology and Immunology
UCLA School of Medicine
Los Angeles, CA 90024

Alkyl phospholipids are diverse chemical species characterized by the presence of an ether bond at the *sn*-1 position of the glycerol moiety. This ether linkage confers unique pharmacological properties to this group of compounds and distinguishes them from the other phospholipids. Alkyl phospholipids are now known to be more abundant in various cell types than previously thought, and there is considerable evidence that certain of these compounds play a crucial and critical role in the regulation of various biological processes including the immune response.

Introduction

Pharmacological interest in alkyl phospholipids developed in the 1960s when Munder *et al.*[1] initiated studies on the physiological role of phospholipid metabolism in macrophages. Subsequently, platelet-activating factor (PAF) was identified as the molecular entity mediating the interaction noted between leukocytes and platelets of immunized rabbits. It was then determined that PAF was a lipid, and semisynthesis of the compound was achieved in the late 1970s. The structure of PAF was subsequently established as a l-O-alkyl-2(R)-

acetyl-glycero-3-phosphocholine,[2-5] an alkyl phospholipid sensitive to phospholipase A_2.[6,7] Confirmation of this structure was obtained by the total synthesis of 1-O-alkyl-2(R)-acetyl-glycero-3-phosphocholine, a compound exhibiting biological activities of natural PAF.[8]

PAF is produced by several inflammatory cells including neutrophils, eosinophils, monocytes/macrophages, platelets, and endothelial cells.[9-21] In addition, many of the cells or tissues, that generate the PAF are themselves targets of PAF-induced bioactivities. The PAF mediator is not stored in the cells but originates from the lipid precursor, 1-O-alkyl-2(R)-acyl-glycero-3-phosphocholine (alkyl-acyl-GPC). It is released in two steps following activation of (1) a phospholipase A_2, which converts alkyl-acyl-GPC to lysoPAF and (2) an acetyltransferase that acetylates the lyso-compound into PAF.[22-26]

PAF is a pleiotropic molecule with many biological activities. PAF is a mediator of anaphylaxis and inflammation and has also been implicated in shock, graft rejection, renal disease, ovoimplantation, and certain disorders of the central nervous system (CNS) (reviewed in References 20 and 26). Although its precise role in these processes awaits clarification, considerable advances have been made after the discovery of specific PAF antagonists (reviewed in Reference 20).

In this review we will consider the role of PAF in the modulation of the complex system of host immune defenses, in particular the direct effect of the PAF on cellular interactions resulting in an immune response.

Cellular Sources of PAF

PAF is produced by a variety of cell types, many of which are involved in the regulation of immune processes (Table 1).

Basophils

Basophil degranulation through an immunoglobulin E (IgE)-dependent mechanism has been shown to be associated with the genera-

Table 1
Cellular Sources of PAF

Basophils and mast cells:	IgE-dependent mechanism, anti IgE
Monocytes/macrophages:	phagocytosis, ionophore A 23187
Polymorphonuclear neutrophils:	phagocytosis, ionophore A 23187, some chemotactic factors
Eosinophils:	C_5a, fMLP, ionophore A 23187
Endothelial cells:	thrombin, interleukin-l, TNF, vasopressin, LTC_4
Large granular lymphocytes:	Fc receptor stimulation

tion of a soluble mediator able to induce the secretion of histamine from platelets.[27-29] In addition, when platelets are added to buffy-coat cell preparations containing basophils and the cell suspension is challenged with the specific antigen, numerous platelet aggregates at the vicinity of degranulated basophils are observed.[29] Generation of PAF from human leukemic basophils stimulated with the ionophore A 23187 has been reported,[30,31] although the amount of the mediator produced appears very low in relation to the high number of basophils present in such cell preparations. In a study carried out on buffy-coat cells from four leukemic patients, the generation of leukotriene (LT) C_4 was correlated with the number of basophils present in the population, whereas the amount of PAF was not (J.P. Marie and J.M. Mencia-Huerta, unpublished observations). Interpretation of the data is, however, difficult because the cell populations used in these various studies were heterogeneous, and the possibility that PAF was actually generated by cell types other than basophils cannot be excluded.

Mast Cells

Initial studies carried out using non-purified rat peritoneal mast cells suggested that this cell type was able to generate the PAF mediator.[9,32] Thus, stimulation of whole peritoneal cell populations with either the ionophore A 23187, compound 48/80, or anaphylatoxins induced PAF release. However, no detectable release of PAF from purified mast cells could be demonstrated under conditions whereby cell functions were preserved and the catabolism of the possibly generated mediator was excluded.[16] Other studies have suggested that serosal mast cells may generate PAF, although under certain experimental conditions the mediator is not released into the medium.[33] However, because these experiments were conducted with mast cell preparations contaminated by other cell types, the possibility remains that the origin of the mediator could have been improperly attributed to mast cells.

Bone marrow cells cultured in the presence of supernatant from either activated T cells or the WEHI 3 cell line differentiate into a homogeneous population of granulated cells.[34-37] These bone marrow-derived cells were later characterized as mast cells on the basis of their functional and biochemical properties.[38-41] These cells store histamine, although in lower amount than serosal mast cells; express high affinity IgE receptors; and contain chondroitin E glycosaminoglycan in their granules. Depending upon the presence of interferon in the medium used for cell differentiation, mouse bone marrow-derived mast cells can express immune response gene-associated antigen on their surface,[42] a function relevant for the modulation of local immunity. Bone marrow-derived mast cells differ from the serosal mast cells in many respects,[43] including the generation of PAF when stimulated with the ionophore A 23187 or after passive sensitization with monoclonal IgE with the specific antigen.[38,44-46]

Polymorphonuclear Neutrophils and Eosinophils

PAF is generated from polymorphonuclear neutrophils (PMN) stimulated with the ionophore A 23187 or by opsonized zymosan.[47,48] The release process is calcium and energy dependent. Human PMN transiently release lysoPAF along with PAF. In addition, the release of PAF is markedly enhanced by lysoPAF, a consequence of the high level of acetyltransferase present in this cell type.[49] PAF exerts various biological effects on PMN that are summarized in Table 2.

Table 2
PAF Effects on Polymorphonuclear Leukocytes

- PAF elicits motile functions as assessed by shape change, aggregation, and random or directed migration.
- PAF induces chemokinesis and chemotaxis.
- PAF induces granule exocytosis.
- PAF is a weak stimulus of the respiratory burst but can be markedly enhanced by other agonists (priming).
- PAF enhances responsiveness of PMN to chemotactic peptides (priming).

Table 3
PAF Effect on Eosinophils

- PAF is a potent chemotactic agent.
- PAF induces LTC_4 and O_2^- formation, and major basic protein release.
- PAF induces cytotoxicity against shistosomes.

Human-like eosinophils produce PAF when stimulated with the ionophore A 23187.[15,50,51] A PMN, this cell type appears to be both a source and a target for the mediator (Table 3). Given the increasing importance attributed to this cell type in asthmatic reactions (reviewed in Reference 20), the generation of PAF by eosinophils may also represent a critical step in the triggering and perpetuation of allergic processes.

Platelets

Rabbit platelets stimulated with the ionophore A 23187, thrombin, or collagen generate PAF.[18] Production of PAF, as well as its action on platelets is independent of the cyclooxygenase- and adenosine diphosphate (ADP)-dependent pathways of activation.[18] Production of PAF is associated with the generation of large amounts of lysoPAF, corroborating the fact that under certain conditions, this inactive precursor may be acetylated by other cell types.

Macrophages/Monocytes

Rat and mouse adherent peritoneal cells generate PAF when stimulated with the ionophore A 23187 or phagocytosable particles, which indicates that macrophages are a source of the mediator.[16,17] The generation of PAF by macrophages is triggered by zymosan particles, opsonized or not; IgG-coated erythrocytes; or immune complexes. However, PAF is not generated by latex beads that are readily ingested indicating that phagocytosis is not sufficient to induce the release. In addition, stimulation of rat adherent cells with zymosan particles in the presence of colchicine or cytochalasin B that block internalization still leads to the generation of PAF, thus indicating a receptor-mediated event.[17] Indeed, the presence of binding sites for the activators of the alternative pathway of complement, such as zymosan particles,[52] which trigger prostaglandins synthesis,[53] has been demonstrated on the macrophage membrane.

The generation of PAF from murine macrophages is dependent on their state of activation. Inflammatory macrophages obtained after intraperitoneal injection of intact bacteria (or their purified extract) release less PAF compared with resting cells.[54-56] Furthermore, macrophages obtained after injection of thioglycolate in the peritoneal cavity release little, if any, of the mediator.[55] Besides releasing less PAF, activated macrophages retain in their membrane (or cell-associated) more of the mediator than they generate.[57] Because these various macrophage populations show similar amounts of alkyl-lipid precursors in their membrane, while they generate lysoPAF in comparable amounts, and exhibit similar acetyltransferase activity, the reduced release of the mediator was attributed to a high rate of catabolism.[55,57]

Alveolar macrophages from rabbit, rat, and humans generate PAF when challenged with the ionophore A 23187, whereas the generation of the mediator during stimulation with zymosan particles appears to be species-dependent.[58] Alveolar macrophages from asthmatic patients have been shown to release PAF following stimulation with the specific antigen.[59] If confirmed, the generation of PAF from

Table 4
PAF Effects on Macrophages

- PAF elicits aggregation and random and directed migration.
- PAF induces chemokinesis and chemotaxis.
- At high concentrations, PAF induces granule exocytosis.
- PAF enhances responsiveness of macrophages to chemotactic peptides (priming).

sensitized alveolar macrophages may represent a new pathway for the development of inflammatory reactions in the lung. This could be critical for pulmonary functions because PAF, similarly to the antigen, has been shown to induce bronchial hyper-reactivity in humans[60] and guinea pigs.[61]

Human monocytes also generate PAF on stimulation with the ionophore A 23187 or zymosan particles.[62] However, in view of the effect of PAF on various immune and macrophage functions (Table 4 and following text), such generation of the mediator by marginated monocytes could affect the local immune processes.

Endothelial Cells

PAF has been shown to increase vascular permeability, both in the skin[63-65] and the lung.[66,67] In lung, this increase in permeability is associated with edema formation and thromboxane A_2 (TxA_2) generation. PAF is also generated from endothelial cells following stimulation with various physiological agonists such as thrombin,[6,12] interleukin 1, and tumor necrosis factor (TNF). PAF acts on this cell type by altering calcium fluxes and morphology,[68-70] and thus could play a major role in the regulation of vascular permeability.[71-73] PAF antagonists such as BN 52021 have been shown to inhibit the effects of the autacoid on endothelial cells, and this action may thus explain part of the prevention of various immunopathological manifestations by this drug.[72,73]

Table 5
PAF and Pathological Manifestations

Acute inflammation	Asthma	Schock
Cardiac anaphylaxis	Digestive ulceration	Psoriasis
Renal disorders	Retinal and corneal diseases	Arterial thrombosis
Immune response		

PAF and Pathological Manifestations

Over the last several years, PAF has been shown to be implicated in various immunopathological situations (Table 5).

Acute Bronchopulmonary Effects of PAF

The specific desensitization of platelets following antigen administration to sensitized rabbits suggested that PAF was generated and led to the hypothesis that the autacoid was a primary mediator in anaphylactic shock.[20] Latter, Vargaftig *et al.*[20,74-77] demonstrated that intravenous injection of natural or synthetic PAF into guinea pig induces a dose-dependent bronchoconstriction related to the effects of the mediator on platelets. This is also confirmed by experiments showing that treatment of guinea pigs with aspirin (which blocks thromboxane synthesis) in combination with mepyramine and methysergide (which inhibit the action of histamine and serotonine released from the activated platelets) abrogates PAF-induced bronchoconstriction.[75,76] In this case, however, the decrease in the number of circulating leukocytes and hypotension, which are unrelated to platelet activation, are still observed.

Interestingly, the pharmacological modulation of the effects of PAF depends on the route of administration of the mediator. As stated previously,[75,76] when the intravenous route is used, blockade of platelet activation or inhibition of synthesis of TxA_2 and action of histamine and serotonin inhibit PAF action. In contrast, when administered by aerosol, PAF evokes a bronchoconstriction that is blocked by aspirin alone, which suggests that the phospholipid recruits primarily cyclooxygenase arachidonic metabolites.

Besides its action on platelets, PAF also exerts direct effects on lung tissues. Injection of the mediator *via* the pulmonary artery of isolated perfused guinea pig lungs induces a potent bronchoconstriction that is blocked by aspirin.[78] Studies by Pretolani *et al.*[79-81] also indicate that the *in vitro* sensitivity of guinea pig lung tissue is influenced by the state of immunization of the animals. Whereas the bronchoconstrictor effect of PAF on lung from control guinea pigs is inhibited by cyclooxygenase blockers,[78] indomethacin has no action on the PAF-induced bronchoconstriction when lung from actively sensitized animals are used. Results from Pretolani *et al.*[80] indicate that leukotrienes play a major role in the PAF-induced bronchoconstriction when the lungs are obtained from sensitized guinea pigs because the dual cyclo- and lipooxygenase blocker BW 755C is inhibitory whereas indomethacin is not. Recent data by Pretolani *et al.*[82] also demonstrate that various PAF antagonists fail to inhibit the bronchoconstrictor and secretory actions of PAF when the mediator is injected into the pulmonary artery of lungs from actively sensitized guinea pigs. This latter result is surprising because the edema formation induced by PAF in those lungs is blocked by these antagonists.[82] These findings raise the possibility that there may exist different subtypes of PAF receptors on the lung tissue (reviewed in Reference 20 and following text) and that their surface expression may vary following immunization.

Antigen challenge of guinea pigs passively sensitized with autologous IgE-rich antiserum for a 10-day period evokes a bronchoconstriction associated with only leukopenia. The PAF antagonists BN 52021[83,84] or WEB 2086 inhibit the antigen-induced bronchoconstriction. In addition, the bronchoconstriction induced by an aerosol of antigen is reduced by BN 52021 or Ro 19-3704. Passive sensitization of guinea pig with heterologous serum from rabbit containing mostly antigen-specific IgG is also frequently used as a model of anaphylactic reactions. In this case, antigen challenge induces a bronchoconstriction associated with thrombocytopenia, neutropenia, and hypotension. The PAF antagonists BN 52021[83] and WEB 2086 markedly reduce the antigen-induced bronchoconstriction. In addition, BN 52021 prevents the antigen-induced thrombocytopenia but is ineffective on the neutropenia.

When animals are actively sensitized with heterologous proteins such as ovalbumin, the bronchopulmonary alterations observed upon antigen challenge depend on the protocol of immunization.[20] When the animals are sensitized using a protocol favoring the production of IgE, poor protection by either BN 52021 or WEB 2086 is afforded (reviewed in Reference 20). In contrast, when the animals are sensitized so as to produce primarily IgG, a 50 to 60 percent reduction of the antigen-induced decrease in dynamic compliance and an increase in pulmonary resistance by BN 52021 are observed. This inhibition is enhanced by pretreatment of the guinea pig with an antihistamine drug (Braquet *et al.*, unpublished results).

PAF and Bronchial Hyperreactivity

One of the major characteristics of asthma is a nonspecific increase in airways responsiveness to various agonists and atmospheric pollutants. It has been suggested that eosinophils are involved in the damage of respiratory epithelium and mucociliary apparatus, and these cells are the prime effector cells in the pathology of asthma. Studies in asthmatic patients have also shown that cell filtration, including mast cells and eosinophils, within the lung parenchyma, is frequently associated with bronchial smooth muscle hyperplasia.

Recently, the involvement of PAF in bronchial hyper-reactivity has been strengthened because administration of the mediator to humans[60] and guinea pigs[61] leads to a long-lasting increase in the responsiveness to metacholine and histamine. Because asthmatic patients are often in repeated contact with the antigen, which in turn may lead to the generation of PAF, the *in vivo* effects of long-term infusion of the phospholipid mediator in the guinea-pig has been investigated.[86] Osmotic Alzet minipumps containing PAF were placed under the skin of guinea pigs and connected to the jugular vein. Animals receiving PAF for two weeks exhibit a nonspecific increase in the response to intravenous injections of histamine as compared with controls. This alteration of lung responsiveness observed following long-term infusion of PAF is associated with changes in lung morphology. Indeed, the lungs of the animals receiving the mediator showed marked hypertrophy of the Reisseisen's muscles associated with muciparous hyperplasia and bronchial obstruction, similar to that observed in lungs from asthmatic patients. PAF has also been shown to increase mucus secretion in the ferret, suggesting that the mediator could have exerted a direct effect on mucus cells.[20] The hyperplasia of bronchial smooth muscle cells may thus explain the hyper-responsiveness observed during histamine provocation of the PAF-treated animals. In addition, eosinophil infiltration and mast cell hyperplasia are observed in lungs from PAF-treated guinea pigs. The latter effect of the mediator is probably related to its action on lymphocytes (see following text), whereas the PAF effect on eosinophils may be a direct one.

PAF in Gram-Negative Bacterial Infection

The lipopolysaccharide constituents of the outer cell wall of gram-negative bacteria are responsible for many of the pathophysiological effects associated with infection with these microorganisms. The intravenous injection of PAF induces effects that are reminscent to those induced by endotoxins, that is, systemic hypotension, pulmonary hypertension, increased vascular permeability, and cardiac dysfunction.[87] Thus, PAF might also play a role in endotoxin-induced pathophysiology, and this has been substantiated in animal studies. The PAF antagonist CV-3988 inhibits the endotoxin-induced hypotension and improves the survival rate after intravenous administration of endotoxin to rats.[88,89] Similar results were reported for BN 52021 treatment in rats[90] and guinea pigs.[91,92]

PAF in Gastrointestinal Alterations

Besides being a potent platelet aggregating agent and a mediator of allergic and inflammatory processes (reviewed in Reference 20). PAF has also been shown to be a potent ulcerogenic agent in the stomach.[93,94] Indeed, intravenous infusion of PAF in the rat induces extensive hemorrhagic damage to the gastrointestinal mucosa characterized by severe hyperemia and mucus hypersecretion. Extensive damage to the surface epithelium is also noted, with focal regions of deeper necrosis. The PAF-induced damages appear more pronounced in the stomach than in the intestine where clear necrosis areas are of limited frequency. Nevertheless, macroscopic alterations of the jejunum and ileum mucosa are observed, whereas the distal colon remained intact, as noted by Wallace and Whittle.[94] This ulcerogenic property of PAF is neither mediated *via* platelet activation or generation of cyclooxygenase products; it is also not mediated *via* stimulation of histamine or adrenergic receptors.[93]

PAF-induced ulcerations are similar to the gastrointestinal impairments observed after endotoxin administration;[93,95,96] like endotoxin, PAF induces hemorrhagic damage associated with vascular congestion in the mucosa of stomach and small intestine but not in that of the distal colon.[94] In addition, PAF also appears to be involved in necrotizing enterocolitis.[97,98]

Several lines of evidence suggest the involvement of PAF in the ulcerations of the stomach and duodenum observed during septic shock : (1) Intravenous infusion of PAF induces many of the symptoms of endotoxic or septic shock, including gastrointestinal ulcerations;[95,96,98] (2) PAF is not present in the blood of control animals but appears in that of rats treated with endotoxin;[99] (3) At doses that inhibit PAF-induced shock, several antagonists, for example, CV 3988,[89] kadsurenone, BN 52021,[91,92] inhibit endotoxin-induced hemodynamic changes in animals. In addition, mediator release and alterations of physiological parameters following ischemia reperfusion in the gastrointestinal tract are markedly reduced upon treatment with various PAF antagonists. For instance, Tagesson *et al.*[100] have investigated how these products influence the damaging effects of ischemia in the small intestinal mucosa in the rat. In their experiments, a ligated loop of the distal ileum was subjected to ischemia/reperfusion, and the ensuing mucosal damage was assessed. Ischemia and reperfusion alone increased the following: mucosal permeability to sodium fluorescein, release of N-acetyl-β–glucosaminidase from the mucosa into the lumen, malondialdehyde content, and myeloperoxidase activity in the mucosa. Intravenous injection of PAF caused a significant potentiation of all these ischemic effects, at doses where it had no damaging effects on the mucosa of animals not subjected to ischemia. PAF antagonists significantly reduced the permeability increase following ischemia and the increases in N-acetyl-β-glucosaminidase release and malondialdehyde accumulation. However, PAF antagonists have no effect on the increase in myeloperoxidase activity following ischemia. These findings suggest a role for PAF in ischemic intestinal injury, possibly by promoting neutrophil-dependent free-radical generation.

In a further study by Droy-Lefaix[101,102] on rat gastric postischemic lesions, the activities of different doses of PAF antagonists were compared with those of superoxide dismutase and allopurinol. Indeed, these two latter compounds are known to provide similar protection against ischemia/reperfusion damage. In rat stomach, 1 hour of ischemia and 24 hours of reperfusion induced a marked mucosal injury involving a drop in adherent mucous gel spinability (a rheological parameter reflecting the state of glycoprotein polymerization), necrotic and hemorrhagic lesion formation into the mucosa, edema formation, and neutrophil invasion into the mucosa, submucosa and muscularis mucosa. Pretreatment with superoxide dismutase, allopurinol, or PAF antagonists gave significant protection against postischemic mucosal damage, the free-radical scavenger and PAF antagonists being equally effective. These results suggest a possible pathological role for both oxygen-derived free radicals and PAF in gastric mucosal ulcerations.

Although PAF is a very potent ulcerogenic agent, it is not involved in all types of ulceration of the gastrointestinal tract. Indeed, Wallace *et al.*[103] demonstrated that intravenous injection of BN 52021 at 10mg/kg fails to prevent ethanol-induced gastric damage in the rat.

As well, PAF does not appear to be involved in gastric hypersecretion in pylorus-ligated rats and in aspirin-induced gastric damage whereas, in contrast with ranitidine, PAF antagonists usually only afford a mild or nil protection regardless of the dose administered. Therefore, the potential therapeutic use of PAF antagonists such as BN 52021 and its analogues in gastrointestinal ulcerations, including those induced by sepsis or stress in man, deserves further investigation.

Graft Rejection

An initial step in rejection is the accumulation of lymphocytes within the graft, which occurs during the first three to five days after transplantation. After this period, the increase in lymphocytes in the graft results mainly from *in situ* proliferation. In cell-mediated rejection, the lymphocyte requires two signals from the macrophage in order to proliferate: foreign antigen presentation and interleukin (IL) 1 production. Once stimulated, the lymphocytes produce IL 2, which is needed for clonal proliferation of helper and cytotoxic T cells. As macrophages are stimulated by PAF to synthesize IL 1, the PAF antagonists may have an effect on lymphocyte proliferation *via* modulation of this process. Such direct or indirect effects of the mediator on IL 1 and IL 2 production may explain the protective action of BN 52021 on graft rejection.[105,106] These studies have shown that treatment of the rat with BN 52021 increases cardiac allograft survival, with the PAF antagonist acting synergistically with azathioprine and cyclosporin A (CSA).

In addition, PAF is released from renal allografts during hyperacute rejection.[107] Recently, Pirotzky *et al.*[108] demonstrated that BN 52021 prevented CSA-induced nephrotoxicity without altering the immunosuppressive effect of the drug. In addition, Pignol *et al.*[109] have shown that treatment with BN 52063 (a ginkgolide mixture of BN 52020, BN 52021, and BN 52022, weight ratio 2:2:1) in combination with CSA increased the immunosupression induced by the latter agent. Both of these results suggested that PAF is generated by

CSA. Whether the *in vivo* effect of BN 52021 and BN 52063 is due to the antagonism of PAF-enhanced cytotoxic cell activity, the generation of suppressor cells, or the action of the PAF antagonists on other cellular or fluid-phase factors involved in graft rejection remains to be defined.

PAF and the Regulation of Cellular Immune Response

Given its potent pro-inflammatory activities, PAF undoubtedly participates in the complex system of host defenses. Recent research efforts, assisted by the development of PAF receptor antagonists and agonists have addressed the potential role played by PAF in modulating cellular immune responses. PAF may modulate these processes either directly by affecting the immunocompetent cells involved or indirectly by influencing the release of other mediators that are operative during the immune response.

Effect of PAF and PAF Antagonists on Cells That Participate in the Immune Response

The Monocyte/Macrophage

Cytotoxicity. Natural cytotoxic cells and macrophages mediate their cytotoxic activities, at least in part, through synthesis and release of TNF. Peritoneal blood-derived monocytes (PBM) cultured overnight with various concentration of PAF (5-30 μM) did not stimulate direct cytotoxicity when tested with ^{51}Cr-labeled target cells. These results suggest that PAF alone has no direct stimulatory activity. Since interferon gamma activates monocytes directly, it was tested whether PAF could enhance this interferon-mediated effect. The results show that a combination of PAF and interferon gamma does not significantly enhance cytotoxic activity.[110]

The supernatants derived from the PBM cultures with PAF were not cytotoxic. However, when the supernatants were tested for the presence of TNF-α using a sensitive radioimmunoassay, even at concentrations of less than 10 μM PAF added to the monocytes, significant production of TNF was detected. These results suggested that PAF can stimulate monocytes to secrete biologically inactive TNF. When PAF and interferon gamma were used together in monocyte cultures, the supernatant contained significant amount of TNF as assessed by radioimmunoassay. These results suggested that additive or synergistic effects are obtained by PAF and interferon.[110] When preincubated, or co-incubated with PAF, LPS-treated human monocytes and mouse peritoneal macrophages also produced significantly higher quantities of TNF.

Secretion of cytokines. A biphasic response is observed when increasing concentrations of PAF are added to LPS-stimulated human monocytes.[111] Low concentrations of PAF (1pM to 0.1nM) significantly enhanced IL 1 production whereas high concentrations (10nM to 1μM) were markedly inhibitory. These effects of PAF on IL 1 production are blocked by the PAF antagonist BN 52021. Under certain experimental conditions, (R)PAF or the PAF analogues PR 1501 and PR 1502 by themselves but not (S)PAF are able to induce IL 1 production, although to a lesser extent than LPS.[112] Recently, the *in vivo* action of PAF on IL 1 production has been confirmed in experiments using osmotic minipumps.[113-114] At low doses (9μg/kg over a seven-day period), PAF induces in increase in IL 1 production by adherent monocytes, whereas at higher doses (28μg/kg over a seven-day period), a decrease is observed. Further, PAF markedly enhances the LPS-induced production of TNF (as discussed above) and IL-6 from both rat macrophages and mouse fibroblasts.

Induction of the receptor for IgE. Various cell types express a low affinity receptor for IgE (FcεRII/CD23), distinct from that described on mast cells and basophils (FcεRI). Evidence for the presence of FcεRII/CD23 has been obtained by the use of IgE-coated erythrocytes, binding studies with radiolabeled IgE, and more recently by the use of monoclonal anti-FcεRII/CD23 receptor antibodies.[115-118] The expression of this receptor is regulated by various cytokines, and particularly interleukin-4.[119,120] Recently, work from our laboratory has demonstrated that PAF is also involved in the regulation of the expression of FcεRII/CD23. Indeed, incubation of monocytes and B lymphocytes but not T lymphocytes with PAF induces a dose-dependent increase in FcεRII/CD23.[121,122] The increased expression of FcεRII/CD23 on human monocytes by PAF appears to be specific because it is inhibited by two PAF antagonists, BN 52021 and BN 50730, whereas addition of anti-IL-4 antiserum is ineffective.[121] In addition, PAF induces FcεRII/CD23 expression on a human eosinophilic cell line (EoL3) and on a human monocytic cell line (U937).[123]

The release of potent pro-inflammatory mediators by IgE-dependent mechanisms suggests that this immunoglobulin class can also contribute to allergic reactions directly by activating FcεRII/CD23-bearing cells. In a recent study, we have investigated the possible alteration of the expression of FcεRII/CD23 on alveolar macrophages from sensitized rats, treated or not treated with the PAF antagonists BN 52021 and BN 50730 and challenged or not challenged with the antigen. Brown-Norway rats were sensitized by three aerosol exposures of ovalbumin, and the expression of FcεRII/CD23 was assessed by flow cytometry after staining alveolar macrophages collected after bronchoalveolar lavage with the BB10 monoclonal antibody.[124] No expression of FcεRII/CD23 on alveolar macrophages from nonsensitized rats, challenged or not challenged with the antigen, was observed. In contrast, a maximum of 74 percent of alveolar macrophages expressed FcεRII/CD23 when collected 24 hours after antigen stimulation by aerosol, compared with 12 percent of the cells following challenge of the rats with a saline solution (Lagente *et al.*, unpublished results). Pretreatment of rats with BN 52021 (10mg/ml aerosol for 30 minutes) or BN 50730 (25mg/kg p.o., 1 hour before) markedly reduced by 82 percent and 95 percent, respectively, the antigen-induced expression of FcεRII/CD23 on alveolar macrophages. These results suggest a role for PAF in the FcεRII/CD23 expression on alveolar macrophages after aerosol administration of antigen to sensitized animals.

This hypothesis was recently confirmed because aerosol administration of PAF (500μg/ml) to rats induced after 24 hours the expression of FcεRII/CD23 on alveolar macrophages, whereas lysoPAF was inactive. *In vitro* as well, PAF induces a concentration- and time-dependent increase in FcεRII/CD23 expression on rat alveolar macrophages, with a maximum at 1μM and after 24 hours. These results obtained in animal models are also strengthened by the data obtained *in vitro* on human monocytes where increased expression by PAF of FcεRII/CD23 is also observed.[121]

Cytotoxic T Lymphocytes and Natural Killer Cells

The precise role of PAF in the generation of cytotoxic lymphocytes *in vitro* is still unclear, although studies with BN 52021 are leading to an improved understanding of the situation. This PAF antagonist potentiates alloantigen recognition in primary and secondary mixed lymphocyte cultures and enhances the generation of cytotoxic lymphocytes *in vitro*.[125] The presence of BN 52021 throughout the duration of primary and secondary mixed lymphocyte cultures has the greatest enhancing effect on the proliferative capacity of the cells, as measured by [^{3}H]thymidine incorporation. Addition of BN 52021 24 hours or more after culture initiation reduces its enhancing effect on lymphocyte proliferation, suggesting that PAF acts at an early stage of this process.

Similar effects are observed in the mixed cultures employed to generate cytotoxic T lymphocytes. The presence of BN 52021 during the entire 72-hour culture period produces an enhanced level of cell-mediated cytotoxicity, whereas the removal of the antagonist up to 24 hours after culture initiation eliminates the effect. Because in these studies the secondary mixed lymphocyte cultures and the bulk cultures used to generate cytotoxic T lymphocytes already contained adequate IL 2 to support the growth of the cells, the potentiating effect of BN 52021 on alloantigen recognition is probably not due to an enhanced production of this cytokine in the antagonist-treated cultures.[125] Farkas *et al.*[126] have examined the effect of the specific PAF antagonist BN 52021 on *in vitro* rat splenic lymphocyte-induced cytotoxicity of Langerhans islets. The protective effect of BN 52021 (300μM) on the islets, as measured by insulin production, is seen when the Langerhans islets are pretreated with BN 52021. This can be regarded as a direct effect of BN 52021 on the target cells, possibly *via* alterations of calcium flux. Using BN 52021-pretreated splenic lymphocytes, the inhibition is significant as compared to control cytotoxicity but slightly lower than that in the above circumstances. In this latter case, these results suggest that the PAF antagonist may have an effect on lymphocyte proliferation. These authors also noted that the protective effect of BN 52021 is dose-dependent because at a concentration of 10μM, inhibition of cytotoxicity is only transient. The results of Farkas *et al.*[126] suggest that PAF may be involved in the cytotoxic effect of splenic lymphocytes, and their findings provide support for the use of PAF antagonists in the prevention of autoimmune diabetes.

An important area is lymphocytes-mediated cytotoxicity against tumor cells. Recent studies have examined the effect of PAF and PAF antagonists on cytotoxic effector cell functions. Natural killer (NK) cell-mediated lysis of the erythroleukemic target cell line K562 is markedly enhanced by picomolar concentrations of PAF.[127] Preincubation of NK cells with PAF prior to culture with K562 target cells, or delayed addition of PAF, also produces enhanced NK activity, an effect that is inhibited by BN 52021. Furthermore, without PAF addition, the NK-mediated cytotoxicity of rat and human lymphocytes against YAC-1 or K562 target cells, respectively, is significantly diminished by BN 52021. In this case, target cell lysis is reduced while the binding of target cells to effector cells is unaffected. In addition, pretreatment of the K562 target cells with the PAF antagonist produces greater inhibition than pretreatment of the effector cells.[128,129]

In addition, large granular lymphocytes, which comprise the effector NK cells responsible for lysis of K562 target cells, can generate PAF under certain conditions.[11] Thus, it appears that endogenous PAF may assist in the regulation of NK cell functions.

T Lymphocytes

When PAF, or the nonhydrolyzable PAF analogue, 2-ethoxy-PAF is added to human peripheral blood lymphocyte cultures stimulated with the mitogens phytohemagglutinin or concanavalin A (Con A), a concentration-dependent inhibition of lymphocyte proliferation is observed.[130] PAF-induced inhibition of lymphocyte proliferation is prevented by BN 52021 and by the cyclooxygenase inhibitor indomethacin, which suggests that prostaglandins may be involved as second messengers.[130] Similar effects of PAF or ethoxy-PAF on the production of IL 2 by human lymphocytes is observed. The suppression of IL 2 production by PAF is also reversed by BN 52021.[125,130]

Because inhibition of lymphocyte proliferation and IL 2 production may be due to the activation of suppressor cells, experiments have been carried out whereby lymphocytes were preincubated with PAF for 3 to 18 hours, washed, and then added to fresh autologous lymphocytes stimulated with mitogens. This coculture assay demonstrated that PAF activates suppressor cells, which subsequently inhibit lymphocyte cultures in the absence of any further contact with PAF.[131,132] This induction of suppressor cells is accompanied by an increase in the CD8+ T/CD4+ T-cell ratio after 18 to 48 hours. The PAF antagonists BN 52021 and WEB-2086 by themselves also generate some suppressor cell activity, although to a much lesser extent than PAF (reviewed in References 127 and 133).

Isolation and purification of specific leukocyte subsets before preincubation with PAF revealed a remarkable spectrum of activities.[127,133] PAF preincubated monocytes exerted suppression *via* an indomethacin-sensitive mechanism, whereas PAF-preincubated CD8+ T cells exerted suppression *via* an indomethacin-resistant process. It is interesting that PAF-preincubated CD4+ T cells markedly enhanced lymphocyte proliferation, and this effect was not blocked by BN 52021. In a different assay system, PAF and two nonhydrolyzable PAF analogues increased the proliferation of IL 2-stimulated human lymphoblasts, whereas some antagonists (CV-3988 and L-652,731, but not WEB 2086 and BN 52021) inhibited such proliferation. Possibly, endogenously produced PAF may be involved in some step(s) of IL 2-induced proliferation of T lymphoblasts, as indicated by the inhibition of this process by the PAF synthesis inhibitor L-648,611.

An important corroboration of some of the *in vitro* observations just outlined has been provided by the *in vivo* instillation of PAF into rats *via* implanted osmotic minipumps.[113] After seven days of instillation of rats with PAF (1µg/kg/for seven days), the recovered splenocytes showed enhanced IL 2 production in response to Con A, an effect inhibited by BN 52021.

T cells are known to regulate IgE production and are therefore believed to play an important role in the pathogenesis of atopic diseases. Peripheral blood T cells from atopic individuals show decreased suppressor cell activity and numbers, a fact that could be accounted for by the generation of various mediators, including PAF, during stimulation with the antigen. Therefore, the effects of PAF on lymphocyte functions, *via* the regulation of IL 2 production could alter not only local immunity but also affect immunoglobulin production (see following copy).

B Lymphocytes

Limited investigations on the possible role of PAF on human B cell is available. Preliminary studies from our laboratory indicate that PAF inhibit the IL-2-induced proliferation of activated B cells. This effect is dose dependent with a maximal effect at 1µM. However, since monocytes are still present in the B-cell population, the possibility that the inhibition is due to the formation of prostaglandins by the former cell type is still open.

Recent studies from our laboratory also indicate that PAF regulates the production of IgE by human B cells. Indeed, a dose-dependent increase in the production of this immunoglobulin is observed when the cells are incubated in the presence of both the autacoid and IL-4 (Dugas *et al.*, unpublished results).

Endothelial Cells

Beside releasing PAF,[6,12] endothelial cells are also target for the mediator (Table 6). The effect of PAF on this cell type may have very

Table 6
Effect of IL-1 or TNF on Endothelial Cells

- Induces the production of prostacyclin (PGI_2) by vascular cells. PGI_2 is a potent antiaggregating agent and vasodilator.
- Increases production of PAF and facilitates adhesion of PMN.
- Induces production of chemotactic cytokines.
- Induces intercellular adhesion molecules ICAM-1, ICAM-2, and leukocyte adhesion molecules (ELAM-1).
- Induces production of various cytokines such as CSF, IL-1, IL-6, PDGF. G-CSF and GM-CSF induce migration and proliferation of endothelial cells.
- IL-1 and TNF alter the functional properties of vascular cells including arachidonate metabolism, thrombogenic properties, leukocyte recruitment, and cytokine production.

important pathological consequences. Indeed, endothelial cells are the prime cells involved in inflammatory reaction. Adhesion of polymorphonuclear cells and platelets to their surface may initiate a cascade of events which leads to the amplication of the pathologic process.[72,73,134,135] This has been initially demonstrated by the works of Bourgain *et al.*[135] and has been reviewed in detail elsewhere.[72,73,134,135]

Eosinophils

PAF is known to modulate eosinophil functions both *in vivo* and *in vitro* (reviewed in Reference 20) and via these processes PAF may indirectly regulate the cellular immune response. In addition to major basic protein (MBP), eosinophil cationic protein (ECP) and eosinophil protein X (EPX) there are two other highly basic proteins contained in eosinophil granules. At relevant *in vivo* concentrations (0.1nM), these proteins induce suppressive effects on peripheral blood mononuclear cells, which suggests a regulatory role for eosinophils in immunological reactions. PAF has been shown to increase eosinophil cytotoxicity *in vitro*.[137] Systematic study of other diseases involving eosinophil infiltration with impairment of lymphocyte functions may allow a better understanding of how PAF and eosinophil activation are linked in the regulation of the lymphocyte response.

Indirect Effects of PAF on the Immune Response

Interaction of PAF With Leukotriene and Prostaglandin Synthesis

Both of these types of mediators have a potent effect on lymphocyte function and thus affect the immune response.[138-141] Evidence implicating PAF in the control of leukotriene and prostaglandin release will be briefly considered. When PAF is infused into isolated guinea pig lungs perfused with cell-free solutions, increased production of TxA_2 and prostacyclin occurs. Airways and arterial pressure are also elevated, and these effects are suppressed by BN 52021 and other PAF antagonists. Bronchoconstriction and the formation of thromboxane are not inhibited by compound FPL 55712, the receptor antagonist of peptidoleukotrienes, at concentrations that suppress the effects of LTD_4.[20] This indicates that different lung sites are involved with release of TxA_2 triggered by peptidoleukotrienes and PAF. Injection of PAF to the pulmonary artery of perfused rat lung induces a dose-dependent vasoconstriction and edema, both phenomena being blocked by diethylcarbamazine, an inhibitor of leukotriene synthesis. High performance liquid chromatography (HPLC) analysis of the lung perfusate revealed the presence of both LTC_4 and LTD_4.[142]

Kidney tissues also release leukotrienes and prostaglandins in response to PAF stimulation. It has been demonstrated that isolated perfused rabbit kidney stimulated with PAF generates prostaglandin E_2 and TxA_2. PAF also induces the release of thromboxane and prostaglandins from primary cultures of human or rat[143,144] glomerular mesangial cells. In addition, it is interesting to note that PAF stimulates the formation of reactive oxygen species from cultured mesangial cells and causes a decrease in the planar surface of the glomeruli, effects that are inhibited by BN 52021 reviewed in Reference 20).

Influence of PAF on Synthesis of Substance P, Growth Hormones, and Neuropeptides

Substance P (SP) is a decapeptide present in the brain and primary sensory fibers of peripheral nerves. SP has been identified in several species, including man (reviewed in Reference 145). SP has been shown to induce leukotriene and histamine release from mast cells and to elicit human monocyte chemotaxis *in vitro* with a median effective concentration (EC_{50}) as low as 0.1pM. Generation of TxA_2, O_2, and H_2O_2 by macrophages activated by *C. parvum* is also stimulated by SP as is lysosomal enzyme release and phagocytosis of yeast particles. This substance may also directly modulate the immune response. *In vitro* it significantly enhances the uptake of [^{3}H]-thymidine and [^{3}H]-leucine by purified human peripheral blood and mouse T lymphocytes. Furthermore, SP significantly enhances by as much as threefold the *in vitro* production of IgA in lymphocytes from spleen and Peyer's patches.[146] PAF has been shown to induce SP release from guinea pig lung preparations, and this may indirectly participate to the regulation of immune processes.

Prolactin (PRL) and growth hormone (GH) may be important in regulating lymphocyte functions because hypophysectomized rats fail to mount an immune response,[147] while treatment of these animals with PRL or GH but not with corticotropin or other pituitary hormones restores their immunological responsiveness. Furthermore, specific binding sites for PRL are present on human lymphocytes. It is interesting that cyclosporine displaces PRL from these sites, and conversely, stimulation of PRL secretion reverses the immunosuppression induced by CsA. The fact that this latter drug may induce PAF production by certain tissues and that PAF is known to induce PRL and GH release from the anterior pituitary lobe of the rat suggests that an interaction between PRL and PAF may exist in the regulation of immune processes.

PAF and PAF Receptor Heterogeneity

In the past few years, several research groups have directed their efforts toward the detailed characterization of the various lipid species generated by various cell types and tissues regardless of their platelet-activating activity (reviewed in References 148 and 149). It is interesting to note that among the various molecules generated at the time of cell stimulation, those molecules exhibiting activity on platelets have been shown to be limited in number and to possess alkyl chains with a length comprised of between 16 and 18 carbons. Increasing or decreasing the length of the alkyl chain will still produce platelet-activating molecules but with markedly decreased specific activity. Although the presence of the ether oxide group at the *sn* 1 position of the glycerol gives maximal activity on platelets, molecules with an ester bound are biologically active and are also generated at the time of PMN stimulation. Some of the molecules that are generated at the time of cell stimulation and exhibit low biological activity on platelets such as 1O-alkyl-2-acetyl-*sn*-glyceryl-3-phosphatydyl ethanolamine have been shown to influence the rate of catabolism of PAF by the acetylhydrolase. Thus, besides the intrinsic specific biological activities of the various alkyl lipids, the one expressed when the mediator is associated with a complex mixture of molecules may have to be determined.

Among the various possibilities that explain the differences observed in the potency of the PAF antagonists may be the expression of heterogeneity at the receptor level (reviewed in References 148 and 149). This latter hypothesis purports that various subtypes of PAF receptors could be present on the same cell type or that different receptors are present on different tissue. Indeed, this latter hypothesis has been recently supported by data from Hwang[150] showing that the binding sites for the autacoid on human platelet and leukocytes are different, as assessed using various unrelated PAF antagonists. In this study, ONO 6240 has been shown to inhibit PAF binding to platelets with a median effective dose eight times lower than that required for human PMN. Also, data from the Institut Henri Beaufour in Les Ulis, France, indicate that BN 52111 (and BN 52115) are more active on this latter cell type, whereas BN 52021 rather inhibits the effect of PAF on platelets, lung, and stomach. Heterogeneity at the receptor level is also indicated by the works of Valone,[151] which demonstrates that the inhibitory effect of BN 52021 and kadsurenone depend on the cell type studied. Works using labeled 52 770 RP and WEB 2086 have demonstrated that the number and affinity of the putative receptor(s) on human platelets for these PAF antagonists depends on the drug.[152,153] Hayashi *et al.*[154] have shown that some PAF analogues are more active than PAF in activating macrophages. However, in this study, it is unclear whether the increased activity of these compounds, as compared with PAF, is related to a lower rate of catabolism. Thus, until the PAF receptor(s) is fully characterized at the molecular level no definite conclusion can be reached, although the already available data strongly suggest that the presence of either receptor subtypes or conformational changes of one single receptor type are likely.

Concluding Remarks

It is now evident that PAF is a mediator whose function is not restricted to inflammatory and allergic reactions. Indeed, accumulating data indicate that this mediator is capable of enhancing, at very low concentrations, the response to other agonists. The delineation of its role in immune processes, therefore, requires further study. Additionally, PAF may be implicated in the fine tuning of immunological reactions, which may explain the broad spectrum of activities that specific PAF antagonists demonstrate in various experimental pathophysiological situations. However, the precise biological activity of the various PAF analogues secreted by various cell types may indicate that specific molecules act on defined targets such as cells or organs. Therefore, despite the abundant data on the pharmacology of this mediator, much work is still required to determine precisely whether PAF is to be considered merely as an autacoid or a regulatory molecule.

References

1. P.G. Munder *et al.*, *Biochemistry* **344**, 310 (1966).
2. J. Benveniste *et al.*, *C. R. Acad. Sci. (Paris)* **289D**, 1037 (1979).
3. M.L. Blank *et al.*, *Biochem. Biophys. Res. Commun.* **90**, 1194 (1979).
4. C.A. Demopoulos, R.N. Pinckard, and D.J. Hanahan, *J. Biol. Chem.* **254**, 9355 (1979).
5. D.J. Hanahan *et al.*, *J. Biol. Chem.* **255**, 5514 (1980).
6. J. Benveniste *et al.*, *Nature (London)* **269**, 170 (1977).
7. J. Benveniste *et al.*, *Fed. Proc.* **41**, 733 (1982).
8. J.J. Godfroid *et al.*, *FEBS Lett.* **116**, 161 (1980).
9. G. Camussi, J.M. Mencia-Huerta, and J. Benveniste, *Immunology* **33**, 523 (1977).
10. G. Camussi *et al.*, *J. Immunol.* **131**, 1802 (1983).
11. F. Malavasi *et al.*, *Proc. Natl. Acad. Sci. USA* **83**, 2443 (1986).
12. G. Camussi *et al*, *J. Immunol.* **131**, 2397 (1983).
13. M.S. Lewis *et al.*, *J. Clin. Invest.* **82**, 2045 (1988).
14. T.M. McIntyre, G.A. Zimmerman, and S.M. Prescott, *Proc. Natl. Acad. Sci. USA* **830**, 2204 (1986).
15. T.C. Lee *et al.*, *Biochem. Biophys. Res. Commun.* **105**, 1303 (1982).
16. J.M. Mencia-Huerta and J. Benveniste, *Eur. J. Immunol.* **9**, 409 (1979).
17. J.M. Mencia-Huerta and J. Benveniste, *Cell Immunol.* **57**, 281 (1981).
18. M. Chignard *et al.*, *Nature (London)* **275**, 799 (1979).
19. P. Braquet *et al.*, in *ISI Atlas of Science: Pharmacology* (publisher, city, 1987), pp. 167-198.
20. P. Braquet *et al.*, *Pharmacol. Rev.* **39**, 97 (1987).
21. G.A. Zimmerman *et al.*, *J. Cell Biol.* **110**, 529 (1990).
22. J.M. Mencia-Huerta *et al.*, *Fed. Proc.* **40**, 1022 (1981).
23. J.M. Mencia-Huerta *et al.*, *Eur. J. Clin. Invest.* **11**, 20 (1981).
24. E. Ninio *et al.*, *Biochim. Biophys. Acta* **710**, 23 (1982).
25. E. Ninio, J.M. Mencia-Huerta, and J. Benveniste, *Biochim. Biophys. Acta* **751**, 298 (1983).
26. F. Snyder, *Proc. Soc. Exp. Biol. Med.* **190**, 125 (1989).
27. P.M. Henson, *Fed. Proc.* **28**, 1721 (1969).
28. P.M. Henson, *J. Exp. Med.* **131**, 287 (1970).
29. J. Benveniste, P.M. Henson, and C.G. Cochrane, *J. Exp. Med.* **136**, 1356 (1972).
30. R.A Lewis *et al.*, *J. Immunol.* **114**, 87 (1975).
31. R.A.F. Clark, J.I. Gallin, and A.P. Kaplan, *J. Allergy Clin. Immunol.* **58**, 623 (1976).
32. J. Benveniste, G. Camussi, and J.M. Mencia-Huerta, *Fed. Proc.* **36**, 1329 (1977).
33. J.A. Denburg *et al.*, *Agents Actions* **14**, 300 (1984).
34. M. Dy *et al.*, *J. Exp. Med.* **153**, 293 (1981).
35. E. Razin, C. Cordon-Cardo, and R.A. Good, *Proc. Natl. Acad. Sci. USA* **78**, 2559 (1981).
36. G. Tertian *et al.*, *J. Immunol.* **127**, 788 (1981).
37. J.W. Schrader *et al.*, *Proc. Natl. Acad. Sci. USA* **78**, 323 (1981).

38. J.M. Mencia-Huerta *et al.*, *J. Immunol.* **131**, 2958 (1983).
39. E. Razin *et al.*, *J. Exp. Med.* **157**, 189 (1983).
40. E. Razin *et al.*, *Proc. Natl. Acad. Sci. USA* **79**, 4665 (1982).
41. E. Razin *et al.*, *J. Biol. Chem.* **257**, 7229 (1982).
42. G.H.W. Wong *et al.*, *Proc. Natl. Acad. Sci. USA* **79**, 6989 (1982).
43. J.M. Mencia-Huerta and M. Benhamou, in *Asthma: Clinical Pharmacology and Therapeutic Progress*, A.B. Kay, Ed. (Blackwell Scientific Pub., London, 1986), pp. 237-250.
44. J.M. Mencia-Huerta *et al.*, *Fed. Proc.* **42**, 6373 (1983).
45. M. Benhamou *et al.*, *J. Immunol.* **136**, 1385 (1986).
46. M. Benhamou *et al.*, *Fed. Proc.* **44**, 2177 (1985).
47. G.Z. Lotner *et al.*, *J. Immunol.* **1240**, 676 (1980).
48. J.M. Lynch *et al.*, *J. Immunol.* **123**, 1219 (1979).
49. F. Alonso *et al.*, *J. Biol. Chem.* **257**, 3376 (1982).
50. T.C. Lee *et al.*, *J. Biol. Chem.* **259**, 5526 (1984).
51. E. Jouvin-Marche *et al.*, *Fed. Proc.* **43**, 1924 (1984).
52. J. Czop, D.T. Fearon, and K.F. Austen, *J. Immunol.* **120**, 1132 (1978).
53. C.A. Rouzer *et al.*, *Proc. Natl. Acad. Sci. USA* **77**, 4279 (1980).
54. R. Roubin, J.M. Mencia-Huerta, and J. Benveniste, *Int. Arch. Appl. Immunol.* **66**, 174 (1981).
55. R. Roubin *et al.*, *J. Immunol.* **129**, 809 (1982).
56. R. Roubin *et al.*, in *Lymphokines*, E. Pick, Ed. (Academic press, New York, 1983), pp. 249-276.
57. A. Dulioust *et al.*, *Biochem. J.* **263**, 165 (1989).
58. B. Arnoux, D. Duval, and J. Benveniste, *Eur. J. Clin. Invest.* **10**, 437 (1980).
59. B. Arnoux *et al.*, *Am. Rev. Respir. Dis.* **125**, 70 (1982).
60 F.M. Cuss, C.M.S. Dixon, and P.J. Barnes, *Lancet* **2**, 189 (1986).
61. L. Mazzoni *et al.*, *J. Physiol.* **365**, 107P (1985).
62. B. Arnoux *et al.*, *Agents Actions* **12**, 713 (1982).
63. C.B. Archer *et al.*, *Br. J. Dermatol.* **110**, 45 (1984).
64. E. Pirotzky *et al.*, *Microcirc. Endothel. Lymph.* **1**, 107 (1984).
65. D.M. Humphrey *et al.*, *Lab. Invest.* **46**, 422 (1982).
66. P.J. Barnes *et al.*, *Br. J. Pharmacol.* **89**, 764P (1986).
67. T.W. Evans *et al.*, *Clin. Sci.* **76**, 479 (1989).
68. H.S. Jacobs, N.R.W. Wickham, and G.M. Vercellotti, *Prostaglandins* **34**, 171 (1987).
69. F. Bussolino *et al.*, *J. Immunol.* **139**, 2439 (1987).
70. A.M. Northover, *Agents Actions* **28**, 142 (1989).
71. F. Breviario *et al.*, *J. Immunol.* **141**, 3391 (1988).
72. P. Braquet *et al.*, *Int. Arch. Allergy Appl. Immunol.* **88**, 88 (1989).
73. P. Braquet *et al.*, *TIPS* **10**, 23 (1989).
74. B.B. Vargaftig *et al.*, *Eur. J. Pharmacol.* **65**, 185 (1980).
75. M. Chignard *et al.*, *Eur. J. Pharmacol.* **78**, 71 (1982).
76. B.B. Vargaftig *et al.*, *Eur. J. Pharmacol.* **65**, 185 (1982).
77. B.B. Vargaftig *et al.*, in *Platelets in Biology and Pathology*, J. Gordon, Ed. (Elsevier, Netherland, 1981), pp. 373-406.
78. J. Lefort, D. Rotilio, and B.B. Vargaftig, *Br. J. Pharmacol.* **82**, 565 (1984).
79. M. Pretolani, J. Lefort, and B.B. Vargaftig, *Am. Rev. Respir. Dis.* **138**, 1572 (1988).
80. M. Pretolani *et al.*, *J. Pharmacol. Exp. Ther.* **248**, 353 (1989).

81. M. Pretolani, P. Ferrer-Lopez, and B.B. Vargaftig, *Biochem. Pharmacol.* **38**, 1373 (1989).

82. M. Pretolani, J. Lefort, and B.B. Vargaftig, *Br. J. Pharmacol.* **97**, 433 (1989).

83. S. Desquand *et al., Eur. J. Pharmacol.* **127**, 83 (1986).

84. V. Lagente *et al., Prostaglandins* **33**, 265 (1987).

85. F. Berti *et al., Pharmacol. Res. Commun.* **18**, 775 (1986).

86. C. Touvay *et al.*, in *New Trends in Lipid Mediator Research,* P. Braquet, Ed. (Karger, Basel, 1988), pp. 26-31.

87. R. Levi, A. Genovese, and R.N. Pinckard, *Biochem. Biophys. Res. Commun.* **161**, 1341 (1989).

88. Z.-I. Terashita, G.L. Stahl, and A.M. Lefer, *J. Cardiovasc. Pharmacol.* **12**, 505 (1988).

89. C. Terashita *et al., Eur. J. Pharmacol.* **109**, 257 (1985).

90. A. Etienne *et al., Agents Actions* **17**, 368 (1985).

91. S. Adnot *et al., Prostaglandins* **32**, 791 (1986).

92. S. Adnot *et al., Pharmacol. Res. Commun.* **18**, 197 (1986).

93. A. Rosam, J.L. Wallace, and B.J.R. Whittle, *Nature (London)* **319**, 54 (1986).

94. J.L. Wallace and B.J.R. Whittle, *Prostaglandins* **32**, 137 (1986).

95. P. Bessin *et al., Eur. J. Pharmacol.* **86**, 403 (1983).

96. P. Bessin *et al.*, in *Platelet-Activating Factor and Related Lipids*, J. Benveniste and B. Arnoux, Eds. (Elseveir, Amsterdam, 1983).

97. F. Gonzalez-Crussi, W. Hsueh, *Am. J. Pathol.* **112**, 127 (1983).

98. W. Hsueh *et al., Eur. J. Pharmacol.* **123**, 79 (1986).

99. P. Braquet and B.B. Vargaftig, *Transplant. Proc.* **18**, 10 (1986).

100. C. Tagesson, M. Lindahl, and T. Otamiri, in *Ginkgolides: Chemistry, Biology, Pharmacology, and Clinical Perspectives*, P. Braquet, Ed. (J.R. Prous Science, Barcelona, 1988), pp. 553-561.

101. L. Pons *et al., Life Sci.* **45**, 533 (1989).

102. M.T. Droy-Lefaix *et al.*, in *Ginkgolides: Chemistry, Biology, Pharmacology, and Clinical Perspectives*, P. Braquet, Ed. (J.R.Prous Science, Barcelona, 1988), pp. 563-574.

103. J.L. Wallace and B.J.R. Whittle, *Br. J. Pharmacol.* **87**, 92P (1986).

104. P. Braquet *et al., Eur. J. Pharmacol.* **150**, 269 (1988).

105. M.L. Foegh, *Transplant. Proc.* **20**, 1260 (1988).

106. M.L. Foegh *et al., Transplantation* **42**, 86 (1986).

107. L. Makowaka *et al, Prostaglandins* **35**, 806 (1988).

108. E. Pirotzky *et al., Transplant. Proc.* **20**, 199 (1988).

109. B. Pignol *et al., Transplant. Proc.* **20**, 259 (1988).

110. B. Bonavida, J.M. Mencia-Huerta, and P. Braquet, *Int. Arch. Allergy Appl. Immunol.* **88**, 157 (1989).

111. B. Pignol *et al., Prostaglandins* **33**, 931(1987).

112. M.L. Barret *et al., Br. J. Pharmacol.* **90**, 113P (1987).

113. B. Pignol *et al.*, in *New Trends in Lipid Mediator Research*, P. Braquet, Ed. (Karger, Basel, 1988), pp. 38-43.

114. B. Pignol *et al., Int. Arch. Allergy Appl. Immunol.* **88**, 161 (1989).

115. J.P. Dessaint and A. Capron, *Triangle* **27**, 95 (1988).

116. G. Delespesse *et al., Int. Arch. Allergy Appl. Immunol.* **88**, 18 (1989).

117. G. Delespesse, H. Hofstetter, and M. Sarfati, *Int. Arch. Allergy Appl. Immunol.* **90** (Suppl. 1), 41 (1989).

118. J. Gordon *et al, Immunol. Today* **10**, 153 (1989).
119. T. Defrance *et al., J. Immunol.* **141**, 2000 (1988).
120. A. O'Garra *et al., Immunol. Today* **9**, 45 (1988).
121. B. Dugas *et al.*, in *Excerpta* (Medica Asia Ltd, Taipei, 1989), in press.
122. J.M. Mencia-Huerta, B. Dugas, and P. Braquet, in *Immunology and Allergy Clinics of North America*, W. Beerman and T.H. Lee, Eds. (W.B. Sanders Company, Philadelphia, 1989).
123. M. Tanaka *et al., Cell Immunol.* **122**, 96 (1989).
124. M. Capron *et al, J. Exp. Med.* **164**, 72 (1986).
125. B.M. Gebhardt *et al., Immunopharmacology* **15**, 11 (1988).
126. G. Farkas *et al., Prostaglandins* **34**, 158 (1987).
127. P. Braquet and M. Rola-Pleszczynski, *Immunol. Today* **8**, 345 (1987).
128. Y. Mandi *et al., Int. Arch. Allergy Appl. Immunol.* **88**, 222 (1989).
129. Y. Mandi *et al., Immunology* **67**, 370 (1989).
130. M. Rola-Pleszczynski *et al., Biochem. Biophys. Res. Commun.* **142**, 754 (1987).
131. M. Rola-Pleszczynski and S. Turcotte, *Prostaglandins* **34**, 148 (1987).
132. M. Rola-Pleszczynski *et al., J. Immunol.* **104**, 3547 (1988).
133. J.M. Mencia-Huerta *et al., Prog. Clin. Biol. Res.* **308**, 441 (1989).
134. P. Braquet *et al. J. Lipid Med.* **10**, (1989).
135. P. Braquet, R. Bourgain, and J.M. Mencia-Huerta, *Semin. Thromb. Hemost.* **15**, 184 (1989).
136. R.H. Bourgain *et al., Prostaglandins* **30**, 185 (1985).
137. A.J. McDonald *et al., J. Allergy Clin. Immunol.* **77**, 227 (1986).
138. T.W. Behrens *et al., J. Immunol.* **143**, 2285 (1989).
139. M. Rola-Pleszczynski, L. Gagnon, and P. Sirois, *Biochem. Biophys. Res. Commun.* **113**, 531 (1983).
140. L. Gagnon *et al., Cell Immunol.* **110**, 243 (1987).
141. M. Rola-Pleszczynski *et al., J. Immunol.* **135**, 4114 (1985).
142. N.F. Voelkel *et al., Science* **218**, 286 (1982).
143. R. Neuwirth *et al., J. Clin. Invest.* **82**, 936 (1988).
144. R. Neuwirth *et al., Circ. Res.* **64**, 1224 (1989).
145. P.J. Barnes, in *Asthma: Clinical Pharmacology and Therapeutic Progress*, A.B. Kay, Ed. (Blackwell Scientific Publications, Oxford, 1990), pp. 58-72.
146. A. Stanisz *et al., Am. Rev. Respir. Dis.* **136**, S48 (1987).
147. E. Nagy *et al., Immunopharmacol.* **6**, 231 (1983).
148. R.N. Pinckard, J.C. Ludwig, and L.M. McManus, in *Inflammation: Basic Principles and Clinical Correlates*, J.I. Gallin, I.M. Goldstein, and R. Snyderman, Eds. (Raven Press, New York, 1988), pp. 139-167.
149. J.M. Mencia-Huerta, D. Hosford, and P. Braquet, in *Allergy and Asthma: New Trends and Approaches to Therapy*, A.B. Kay, Ed. (Blackwell Scientific Publications, Oxford, 1989), pp. 50-68.
150. S.B. Hwang, *J. Biol. Chem.* **263**, 3225 (1988).
151. F.H. Valone, *J. Immunol.* **140**, 2389 (1988).
152. G. Dent *et al., Eur. J. Pharmacol.* **169**, 313 (1989).
153. D. Ukena *et al., FEBS Lett.* **2280**, 285 (1988).
154. H. Hayashi *et al., J. Biochem. (Tokyo)* **97**, 1737 (1985).

7

Lung Preservation:
A New Indication for a PAF Antagonist, BN 52021

John V. Conte
Department of Surgery
Georgetown University

Peter W. Ramwell
Department of Physiology
Georgetown University

correspondence
Marie L. Foegh
Department of Surgery
Georgetown University
Medical Center
Washington, D.C. 20007

Lung preservation is limited by anoxic injury during ischemic storage and reperfusion injury caused by oxygen-derived free radicals and vasoactive substances released during reperfusion. There is a similarity between the pathophysiology of the pulmonary dysfunction following lung preservation and that caused by platelet-activating factor (PAF). Furthermore PAF augments free radical formation and thus may promote lung injury during reperfusion injury as well as during harvesting and preservation. We have found a PAF antagonist, BN 52021, to improve lung transplantation following long term lung preservation.

Lung transplantation as a treatment for end-stage cardiopulmonary and pulmonary disease increases every year and peaked in 1988. Lung transplantation is performed as single, double, and heart-lung allografts, the latter still being the most common modality. The clinical advances made in the field of lung transplantation have occurred largely due to better surgical techniques, improved immunosuppression, and earlier and more reliable methods of detecting and treating episodes of rejection.[1] Despite advances in these areas, lung preservation techniques have failed to progress at the same pace, and organ availability is the major factor limiting access to lung transplantation.[2]

Lung Preservation

Currently, the maximal period of lung preservation for clinical use is four to six hours. Lung preservation is limited by ischemic damage that occurs during the preservation period and by reperfusion injury to the ischemic lung. Pulmonary dysfunction following lung preservation and transplantation is characterized by hypoxia, increased pulmonary vascular resistance, increased lung weight, decreased lung compliance, increased shunting, and pulmonary edema.

The optimal method of lung preservation is not known. However, experimental and clinical studies have shown that lung preservation can be improved by several factors. The commonly used preservation solutions have an intracellular electrolyte composition (that is, high potassium, high magnesium) and offer more protection than solutions with an extracellular electrolyte composition (that is, high sodium, high chloride). Solutions made hyperosmolar with sugars and colloids have improved organ preservation in laboratory studies compared with solutions that are isosmolar or hyposmolar. Substrate enhancement with sugars and insulin are also helpful and are thought to maintain adenosine triphosphate stores. Both the University of Wisconsin and the Eurocollins organ preservation solutions have these characteristics. These solutions are used extensively both clinically and experimentally for organ preservation. Additional protection is offered in lung preservation by drugs that reduce the generation of oxygen-derived free radicals. Included in this latter group are superoxide dismutase, catalase, allopurinol, mannitol, dimethyl sulfoxide, and deferoxamine.[3] The biochemical site of activity of these free radical scavengers is shown in Figure 1.

Platelet-Activating Factor and Lung Preservation

Platelet-activating factor (PAF), 1-alkyl-2(R)-acetyl-glycero-3-phosphoryl-choline is a phospholipid released by different cells that has a variety of pathophysiological effects. In experimental animal models, PAF has been shown to cause bronchoconstriction,[4,5] high

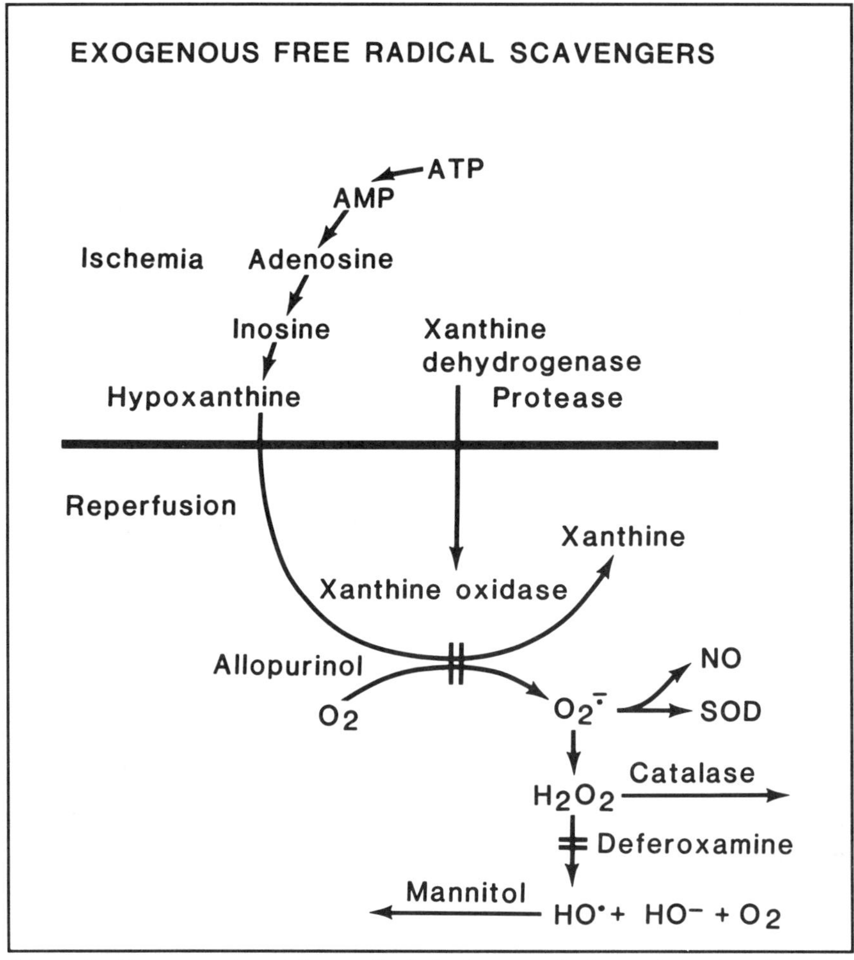

Figure 1. The generation of free radicals during reperfusion and the site of activity of free radical scavengers.

pulmonary vascular resistance,[6] pulmonary edema,[7] and increased bronchial secretions.[7] The mechanism of these effects may be due to PAF-induced generation of vasoactive mediators and oxygen-free radicals.[8]

PAF is released from leukocytes, endothelial cells, macrophages, and other cells. Aspirates obtained from lungs during hypothermic storage show large numbers of polymorphonuclear leukocytes (PMN) and twice as many macrophages. These cells and the endothe-

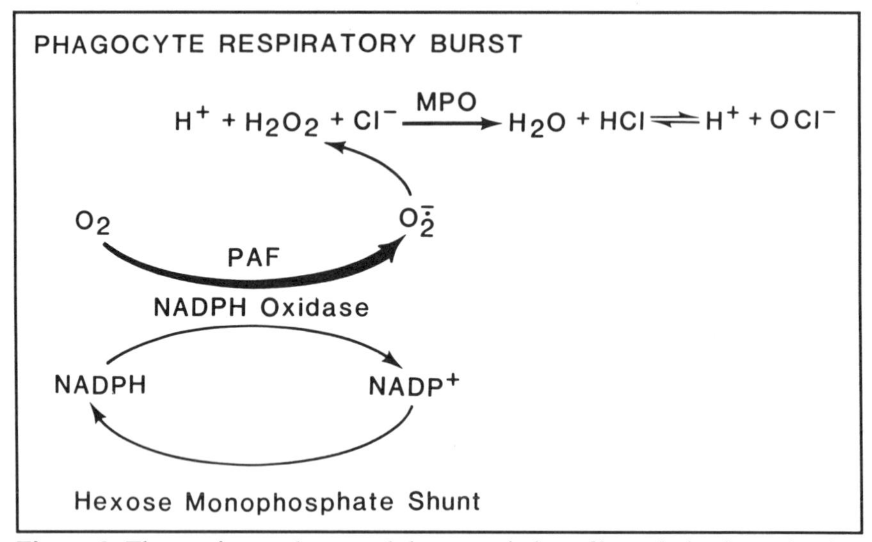

Figure 2. The respiratory burst and the potentiating effect of platelet activating factor (PAF).

lial lining may become activated during reperfusion and release PAF, oxygen-derived free radicals, and other vasoactive substances. PAF has been shown to augment the oxidative burst and thus free radical formation from neutrophils (Figure 2). PAF may have similar effects in other cells like macrophages, which are numerous in lungs. These findings, coupled with the similarity between the pathophysiological effects of PAF and post transplantation pulmonary dysfunction, led to an investigation of the effects of a platelet-activating factor antagonist, BN 52021, on post transplantation pulmonary function in dogs.[9]

Left single lung transplantation[10] was performed in dogs, and the lungs were preseved for 22 hours at 10°C. At the time of harvest, the lungs were flushed with a Eurocollins preservation solution and stored in the same solution. After completion of the transplantation and a 30-minute reperfusion period the native (nontransplanted) pulmonary artery was ligated, and the function of the transplanted lung was followed for six hours. The platelet-activating factor antagonist, BN 52021 (Institut Henri Beaufour, Le Plessis Robinson, France) was given intravenously both to the donor and the recipient. In addition, the preservation solution (Eurocollins) contained BN 52021,

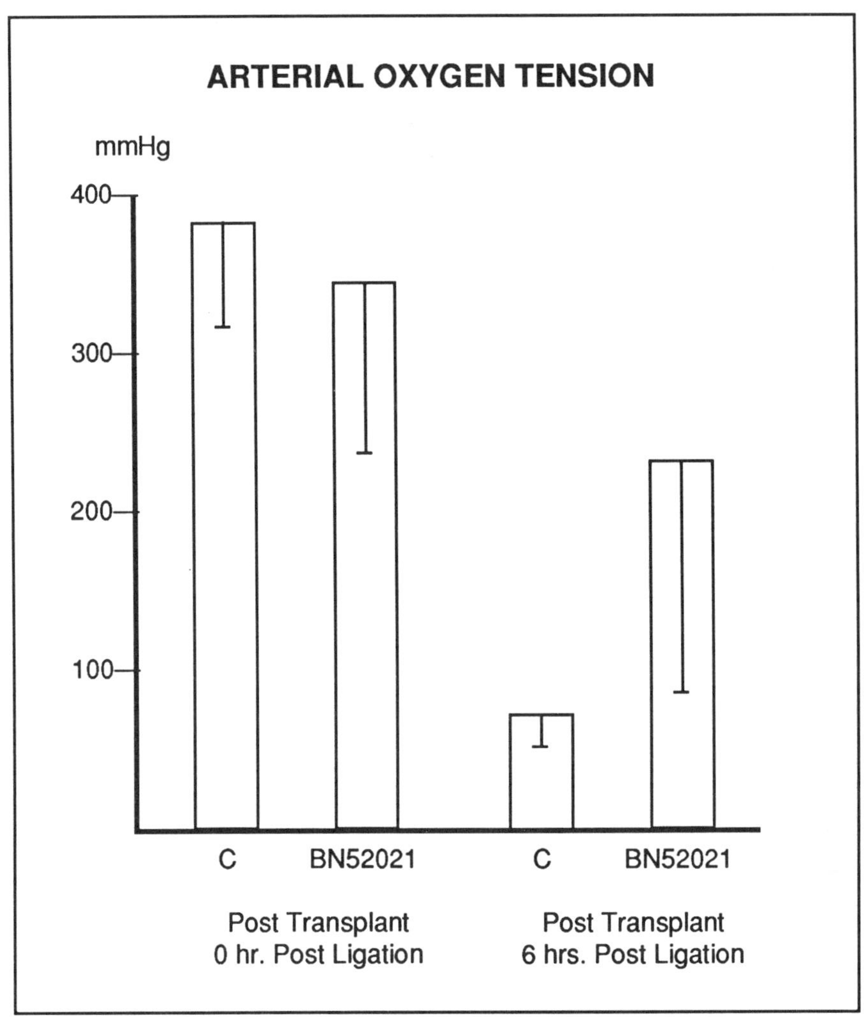

Figure 3. Arterial oxygen tension is shown in the control group (C), and the group receiving the PAF antagonist (BN 52021) at the time of ligation of the pulmonary artery to the native lung (0 hr. post ligation) and at 6 hrs. post transplantation. The lung transplantation is performed following 22 hours of preservation. The BN 52021 treated group has significantly improved oxygen tension compared to the control group. The animals are breathing pure oxygen. All values are mean ± SD.

and the recipient received the BN 52021 30 minutes prior to reperfusion. The dogs in the control group received an equal volume of saline. The pulmonary function was evaluated by following arterial oxygen tension, alveolar oxygen difference, and pulmonary vascular resis-

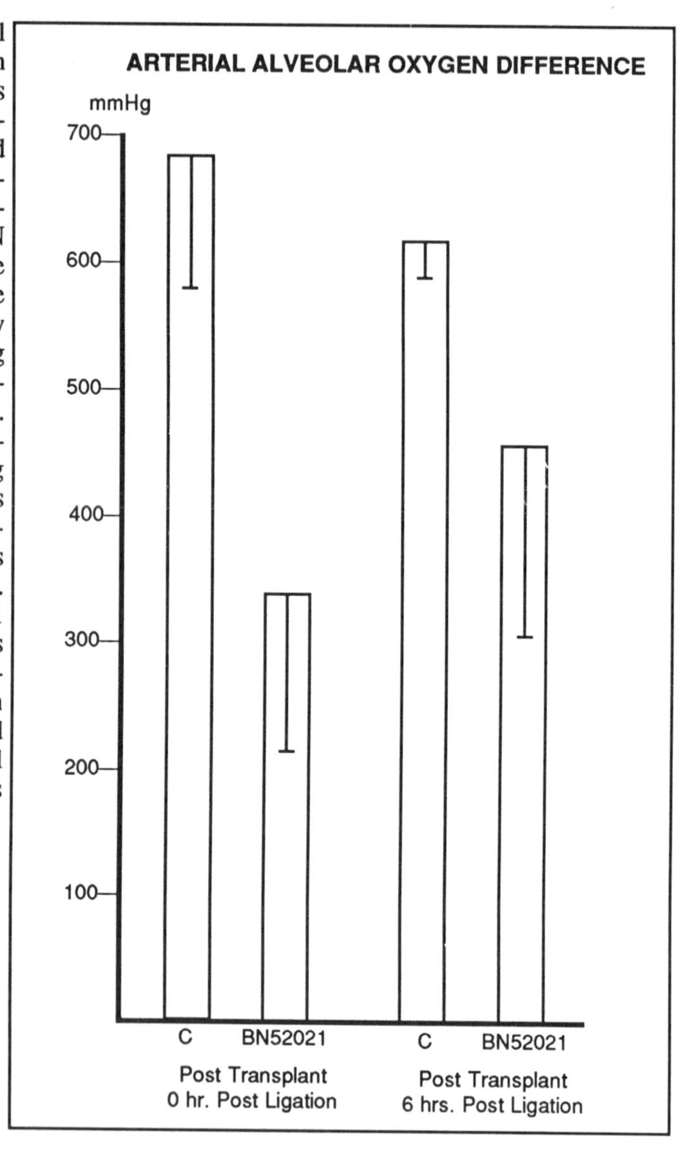

Figure 4. Arterial alveolar oxygen difference is shown in the control group (C), and the group receiving the PAF antagonist (BN 52021) at the time of ligation of the pulmonary artery to the native lung (0 hr. post ligation) and at 6 hrs. post transplantation. The lung transplantation is performed following 22 hours of preservation. The BN 52021 treated group has significantly improved oxygen tension compared to the control group. All values are mean ± SD.

tance. These parameters were obtained at baseline, post pneumonectomy, post transplantation, and hourly for six hours following ligation of the nontransplanted pulmonary artery. The arterial oxygen tension, which is the most reliable indicator of lung function, was consistently

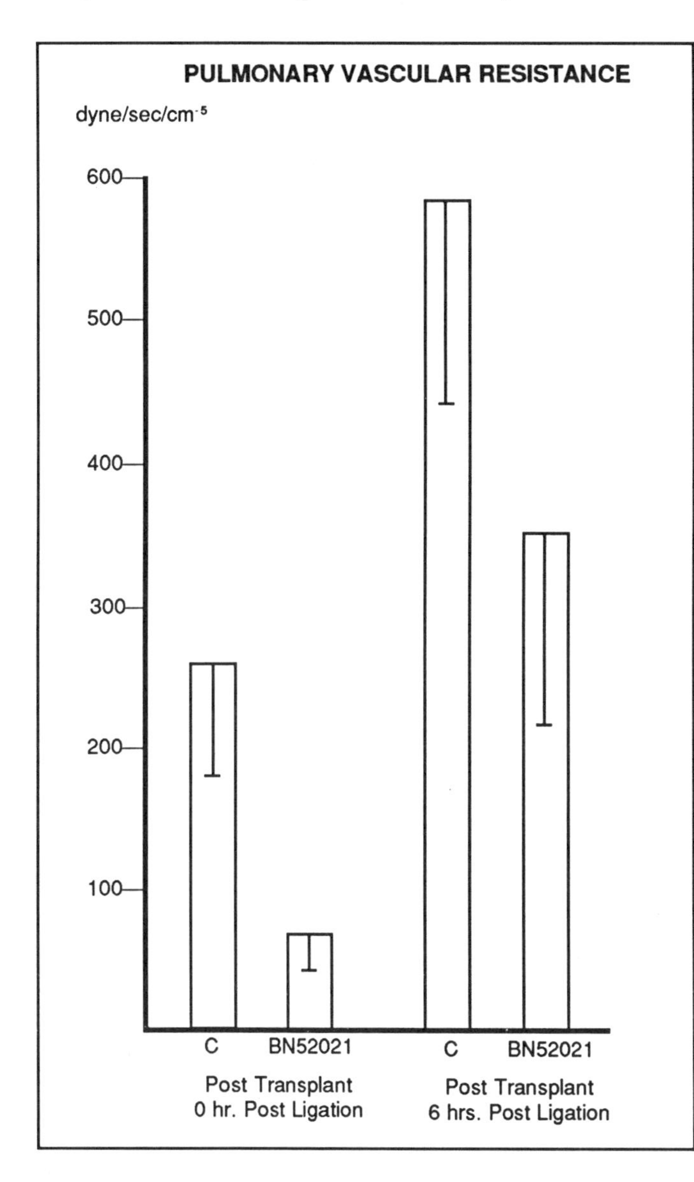

Figure 5. Vascular pulmonary resistance is shown in the control group (C), and the group receiving the PAF antagonist (BN 52021) at the time of ligation of the pulmonary artery to the native lung (0 hr. post ligation) and at 6 hrs. post transplantation. The lung transplantation is performed following 22 hours of preservation. The BN 52021 treated group has significantly improved oxygen tension compared to the control group. All values are mean ± SD.

better in the BN 52021-treated group throughout the observation period. The alveolar arterial oxygen difference at the end of the study was much better in the BN 52021-treated group than in the saline-treated group. The BN 52021-treated group also maintained a lower pulmonary vascular resistance. The data at time of ligation of the pulmonary artery to the nontransplanted lung (0 hours) and six hours post ligation are shown in Figures 3 to 5. There was no difference in these parameters at baseline between the groups.

The PAF antagonist BN 52021 improves lung function following prolonged periods of lung preservation, whereas Eurocollins solution by itself preserves the lungs less well. The additional treatment of the donor may have caused less leukocyte and platelet adhesion to the vasculature during flushing of the lungs. The treatment of the donor prior to reperfusion may protect and attenuate free radical formation from both circulating leukocytes and alveolar macrophages. The better pulmonary function in the dogs receiving BN 52021 suggests these mechanisms.

A further interesting aspect of PAF antagonists in lung transplantation is the synergistic effects with cyclosporin A, an immunosuppressant agent, on graft survival. However, BN 52021 is only a weak immunosuppressant agent in that the drug by itself has no effect on graft survival.[11,12]

Conclusion

Lung preservation continues to be a problem and is a limiting factor in the availability of lung transplantation. Current methods of lung preservation have been improved with the addition of oxygen-free radical scavengers, but the lungs are still only preserved for four to six hours. The study mentioned here shows that the platelet-activating factor inhibitor BN 52021, when administered to donor and recipient, and present in the preservation solution, helps to improve postoperative lung function. Thus, the addition of BN 52021 to a lung preservation protocol may help maintain lung function and possibly extend the allowable periods of lung preservation, making lung transplantation available to more patients.

References

1. G.A. Patterson and J.D. Cooper, *Surg. Clin. N. Am.* **68**, 545 (1988).
2. A. Haverich, W.C. Scott, and S.W. Jamieson, *Heart Transplantation* **4**, 234 (1985).
3. J.V. Conte *et al., J. Thorac. Cardiovasc. Surg.*, submitted.
4. B.B. Vargaftig *et al., Eur. J. Pharmacol.* **65**, 195 (1980).
5. A. Denjean *et al., J. Appl. Physiol.* **55**, 799 (1983).
6. L.M. McManus *et al., J. Immunol.* **125**, 2919 (1980).
7. D.A. Handley *et al., Thromb. Haemost.* **52**, 34 (1984).
8. P. Braquet and B.B. Vargaftig, *Transplant. Proc.* **18**(5), 10 (1986).
9. J.V. Conte *et al., Transplantation*, submitted.
10. F.J. Veith *et al., J. Thorac. Cardiovasc. Surg.* **72**, 97 (1976).
11. M.L. Foegh *et al., Adv. Prostaglandin Thromboxane Leukotriene Res.* **19**, 377 (1989).
12. M.L. Foegh *et al., Transplantation* **42**, 86 (1986).

8

Platelet-Activating Factor:
The Alpha and Omega of Reproductive Biology

correspondence
John M. Johnston
Shuishi Miyaura
Departments of Biochemistry,
Obstetrics-Gynecology, and
The Cecil H. and
Ida Green Center for
Reproductive Biology Sciences
The University of Texas
Southwestern Medical
Center at Dallas
5323 Harry Hines Boulevard
Dallas, TX 75235-9038

The major investigative efforts in the early 70s and mid-70s seventies were focused on the role of platelet-activating factor (PAF) in the inflammatory response.[1] Following the discovery of the structure of PAF in 1979 by Demopoulos, Pinckard, and Hanahan;[2] Snyder and colleagues[3] and Benveniste and coworkers,[4] there has been tremendous growth in both the number of biological systems investigated and the number of publications relating to the role of PAF in inflammation. One such area is reproductive biology. In this chapter, the role of PAF will be discussed, with major emphasis on its role in implantation, fetal development (emphasis on the lung), and the initiation and maintenance of parturition.

PAF in Early Development

A major area of investigation has been focused on the role of platelet-activating factor (PAF) and implantation; however, some observations have been reported concerning PAF in relation to ovulation, sperm motility, and so forth. Although the precise role that PAF may play in ovulation is not completely clear, the addition of a PAF antagonist into the bursa of the ovary has resulted in a marked reduction of the ovulation number in the rat.[5] These investigators[5] also reported that they were able to reverse the inhibition by administering PAF. The concentration of PAF in the ovary has also been determined in animals that had been superovulated.[6] A decrease in PAF was found following human chorionic gonadotropin administration. Based on these early observations, it would appear that PAF, along with other autacoids, is involved in the process of ovulation. A role of PAF in the appearance of early pregnancy factor in serum has also been suggested by the experiments of Orozco *et al.*[7] These investigators demonstrated that the administration of PAF to mice in estrus resulted in the appearance of early pregnancy factor in serum. Previous studies have established that the appearance of this protein in serum occurred within three hours following fertilization in certain species.[8]

A role of PAF in the metabolism of sperm was suggested by the demonstration that PAF is present in the ejaculum of both animals[9] and humans.[10] The motility of so called "sluggish" human sperm was increased significantly by the addition of PAF to the medium.[11]

It has also been proposed that PAF participates in events associated with maternal recognition of pregnancy and implantation. A mild systemic thrombocytopenia is one of the earliest signs of maternal response to fertilization.[12] The presence of a viable embryo is thought to cause the peripheral blood cell count to decrease in both mice[12] and humans.[13] Until recently, it has been difficult to monitor very early pregnancies from the time of fertilization until implantation. In the past, this phase of development has not been well documented. With the development of *in vitro* fertilization procedures, information concerning the metabolism of embryos through the blastocyst stage can be obtained. Primarily due to the investigations of O'Neill,[14] we now have new insights about the role of PAF in implantation. The first evidence that the early embryo produced PAF was an outgrowth of the observation that mice exhibit a splenic contraction and a mild thrombocytopenia shortly after fertilization and prior to the implantation phase of pregnancy. PAF was identified as the factor that induced the thrombocytopenia, and it was demonstrated that embryos produced PAF.[15] It was suggested that during *in vitro* fertilization and embryo transfer, those embryos that did not secrete PAF into the media had a decreased incidence of successful implantation when compared with those embryos that produced significant amounts of PAF.[16] Subsequently, Amiel *et al.*[17] reported that they were unable to detect the presence of PAF in the media obtained from human embryos assayed by a platelet aggregation procedure. PAF was detected, however, by the injection of both the culture media and embryos into splenectomized mice in a manner similar to that previously shown by O'Neill.[15] We have carried out similar studies and were able to detect the presence of PAF in the culture media obtained from human embryos by a serotonin release assay (A. Vereecken, M. Angle, and J.M. Johnston, unpublished observations). We found no difference in the PAF secreted into media by embryos that ultimately developed into viable fetuses when compared with the PAF from embryos that failed to implant. Although these findings appear to be somewhat different from that reported by O'Neill *et al.*,[16] if one considers only the group of patients who became pregnant, the concentration of PAF in the media was significantly higher in those patients who went to term with an uneventful pregnancy compared with those patients whose pregnancies resulted in a spontaneous abortion.

In support of the role of PAF in implantation, O'Neill and colleagues[18] have supplemented the culture media with PAF and determined the successful implantation rate and pregnancy potential of these preembryos produced by *in vitro* fertilization. In a group of 185 women who received either embryos or PAF-treated embryos, a small but significant difference in the pregnancy rate achieved was evident in the PAF-treated group. Based on these findings, it would appear that PAF produced by the developing embryo may have a functional role in the phenomena of implantation. Whether or not the small amounts of PAF produced by the embryo are directly responsible for the maternal thrombocytopenia is not so clear. As an alternative, it has been suggested that the embryonic production of PAF results in the activation of the endometrium of the uterus to produce PAF, thereby further amplifying the signal of embryo PAF production.[19] PAF has been detected in uterine tissue of rat,[20] rabbits,[21] and human.[22] When the PAF concentration was determined in the rabbit uterus during pregnancy, it was found that an increase in PAF was observed from Day 3 of pregnancy or in pseudopregnancy. The PAF concentration in the total rabbit uterus appears to decrease to levels prior to pregnancy by Day 7.[21] Support of the contention that it is indeed the endometrium that is involved in PAF metabolism is the observation that the PAF concentration in the endometrium was significantly higher than that found in the myometrium of the rabbit. It has not been clearly established whether the increase in endometrial PAF is due to an increased synthesis or a decreased breakdown in this tissue. It has been suggested that the capacity to synthesize PAF is relatively high in human endometrial tissue as judged by PAF acetyltransferase activity.[23] The presence of 1-alkyl-2-acetyl-*sn*-glycerol:cytidine diphosphate-choline choline phosphotransferase (a reflection of the *de novo* pathway for PAF biosynthesis) has also been demonstrated in the microsomal fraction obtained from endometrial tissue. Harper and colleagues[24] have reported that PAF is closely associated with the stromal cells of the endometrium.[24] The addition of PAF to glandular cells in culture caused the increase release of prostaglandin E_2 (PGE_2) but not of $PGF_{2\alpha}$.[25] A similar specificity in the stimulation of PGE_2 formation by PAF in amnion cells has also been reported.[26] The presence of PAF receptors in the rat uterus as well as the presence of the PAF receptor in a purified endometrial membrane preparation have recently been demonstrated.[27] Zhu *et al.*[28] also demonstrated the presence of PAF receptor in rat and human myometrium. The mechanism by which PAF binding is linked to the second messenger system in the uterus is not clearly established.

Prostaglandins also appear to be associated with the process of implantation in several species. Treatment with indomethacin in animals has been shown to block the implantation of embryos in mouse[29,30] and rat,[31,32] and the addition of prostaglandins will reverse the inhibition.[29-31]

The concepts that have been discussed in relation to PAF synthesis by the developing fetus and the involvement of PAF in the process of implantation are summarized in Figure 1. At the top of the figure is an illustration of the fertilized ovum developing through the morula into the blastocyst stage and the beginning of the implantation phenomena. The site of implantation is illustrated in the center section

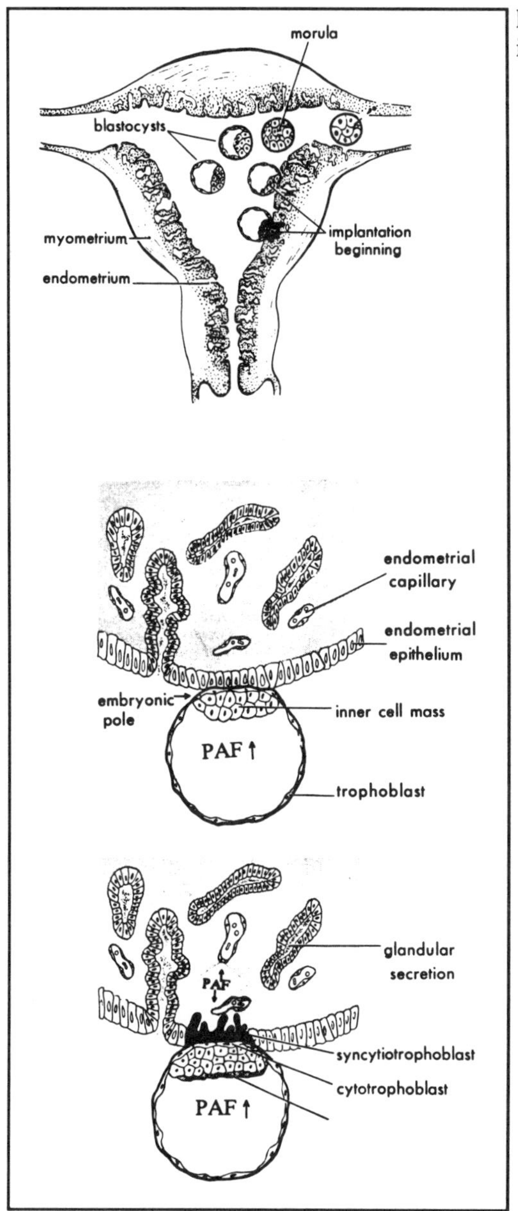

Figure 1. Proposed role of PAF in implantation.

of this figure, in which the preembryo is attached to the endometrial epithelium. Finally, the effect of PAF in implantation is illustrated at the bottom of the figure, with the cytotrophoblast showing the increase in PAF biosynthesis that leads to an increase in PAF production by the endometrium. This process results in the formation of a site during implantation that could be compared with an inflammatory site.

The Role of PAF and Other Autacoids in The Initiation of Parturition

Little information is available concerning the developmental profiles of PAF and the enzymes responsible for its metabolism during the course of gestation. Recently, however, new and important information concerning the role of PAF and eicosanoids as in relation to the onset of labor has been published. The concept that the fetus participates in the process of labor is indeed an old one that was first suggested by Hippocrates (460 to 370 B.C.) who hypothesized that the fetus "becomes agitated and breaks through the membranes."[33]

The communication from the fetus to the maternal compartment is most likely mediated via the amniotic fluid and the amnion tissue. This tissue is morphologically attractive for the transduction of the signal(s) to the chorion laeve and decidua and ultimately to the myometrium and cervix, because (1) it contains an abundance of lipid droplets (the importance of lipids in parturition will be discussed); (2) it contains an abundance of microvilli, intercellular junctions, and cytoplasmic filaments to accept a signal and; (3) amnion tissue has a high activity of various enzymes involved in PAF and eicosanoid biosynthesis, some of which are increased at term.

Our initial interest in the molecular mechanisms involved in parturition developed from the observation that the concentration of unesterified arachidonic acid in amniotic fluid increases disproportionately (six- to tenfold) during labor.[34] Based on these findings it was suggested that a relative specificity for the mobilization of arachidonic acid during labor occurred. It is well established that relatively large amounts of prostaglandins are produced by uterine and intrauterine tissues during parturition (for review, see Reference 35).

The importance of lipids in the initiation of parturition was first suggested by Luukkainen and Csapo[36] who reported that the intravenous infusion of lipid emulsions into pregnant rabbits resulted in an increased responsiveness of the rabbit uteri to oxytocin. Subsequently, it was shown that the active component in these emulsions was a glycerophospholipid containing an essential fatty acid.[37-39] It was also reported that the intravenous administration of arachidonic acid resulted in parturition in rabbits[40] and induced the premature oviposition in quail.[41] The instillation of an arachidonate-albumin complex into the amniotic sac resulted in the termination of pregnancy.[34]

Arachidonic acid must be released from its esterified form in order to serve as a precursor of eicosanoids.[42,43] Karim[44] reported that prostaglandins were present in the amniotic fluid and that prostaglandin-enriched extracts of amniotic fluid obtained from women in labor increased the intensity of spontaneous contractions of the uterus. It was suggested that the arachidonic acid that was found in increased amounts during labor was derived from the glycerophospholipids of fetal membranes.[34] Approximately 20 percent of the total fatty acid content of the amnio-chorion was found to be arachidonic acid. The loss of arachidonic acid from amnion tissue was confined to two glycerophospholipids, namely (diacyl)phosphatidylethanolamine and phosphatidylinositol.[45]

A model for the release of arachidonic acid from glycerophospholipids that occurs during early labor in amnionic tissue was proposed to explain both the loss of this fatty acid and the enzymatic specificity for its release (Figure 2).[46] The presence of an enzyme(s) in fetal membranes that preferentially cleaved phosphatidylethanolamine-containing esterified arachidonic acid in the *sn*-2 position (Rx 1) as well as a phosphatidylinositol-specific phospholipase C (Rx 2) was reported.[46,47] Both enzymes were activated by Ca^{2+} ion. Diacylglycerol (Rx 3) and monoacylglycerol (Rx 4) lipases were also present in various intrauterine tissues.[47] A similar reaction sequence has recently been demonstrated in human platelets by Majerus and colleagues.[48] The diacylglycerols produced in amnionic tissue from the hydrolysis of phosphatidylinositol can also be phosphorylated by

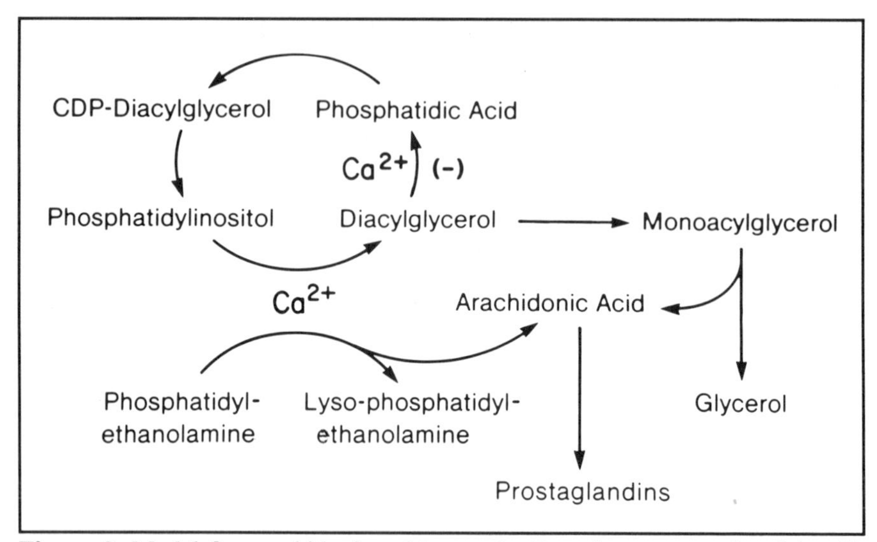

Figure 2. Model for arachidonic acid cascade. Reactions are catalyzed by (1) phospholipase A_2, (2) phosphatidyl-inositol-specific phospholipase C, (3) diacylglycerol lipase, (4) monoacylgycerol lipase, (5) diacylglycerol kinase (6) prostaglandin synthase complex, (7) cytidine triphosphate:phosphatidate cytidylyltransferase, cytidine diphosphate-diacyl-glycerol:inositol 3-phosphatidyltransferase.

a diacylglycerol kinase (Rx 5), forming phosphatidic acid.[49] Diacylglycerol kinase was inhibited by Ca^{2+}. It was suggested that an increase in Ca^{2+} concentration favored the formation of arachidonic acid rather than the formation of phosphatidic acid by this mechanism. This model provides the enzymatic basis for the selective release of arachidonic acid from the two glycerophospholipids and emphasizes a central role for Ca^{2+} in the regulation of the release of this polyunsaturated fatty acid.

The role of Ca^{2+} in the regulation of arachidonate release directed us to examine the role of PAF in parturition because this autacoid is associated with an increase in intracellular Ca^{2+} in a number of cells.[50] PAF was identified and characterized in the amniotic fluid obtained from women in active labor.[51] Platelet aggregation occurred in a dose-dependent manner in response to the PAF purified from amniotic fluid obtained from women at term and in active labor. Only a trace activity of PAF was observed when the amniotic fluid obtained from women at term and not in labor was assayed. These observations have been confirmed by Nishihira and colleagues.[52]

PAF as well as a Ca^{2+} ionophore caused a marked stimulation of PGE_2 formation in amnionic tissue.[53] A similar increase in PGE_2 formation has recently been reported by Bleasdale and colleagues[54] who used amnionic cells in culture. We demonstrated that amnionic tissue has the capacity to synthesize PAF, and the biosynthesis of PAF by the "remodeling" pathway was investigated in this tissue.[53] Maximum activity of PAF acetyltransferase was obtained at low concentrations of Ca^{2+} (0.1 to 1μM). The specific activities of PAF acetyltransferase and acetylhydrolase in fetal membranes were not significantly different throughout gestation or from women at term. The regulation of PAF acetyltransferase by Ca^{2+} has been suggested to occur via Ca^{2+}-dependent protein kinase C;[55,56] this activity is present in amnionic tissue.[57] In contrast, Sanchez-Crespo and colleagues[58] have suggested that PAF acetyltransferase is activated by a cyclic adenosine monophosphate-dependent phosphorylation.

PAF: Its Role in Fetal Lung Maturation

Although amnionic tissue contains the enzymes and substrates that are required for the synthesis of PAF via the remodeling pathway, we were unable to demonstrate a significant release of PAF into the incubation media.[51] Similar findings have been reported for endothelial cells.[59] Amniotic fluid, at term, is enriched with surfactant, a lipoprotein complex that is synthesized by the type II pneumonocytes of the fetal lung and transported to the amniotic fluid in the form of lamellar bodies. The precursor to PAF, namely alkyl-acylglycerophosphocholine, has been shown to be present in surfactant.[60] When the distribution of PAF between the lung lamellar body and supernatant fractions was determined in amniotic fluid obtained from women in labor at term, approximately 44 percent of the PAF was associated with the lamellar body-enriched fraction.[51] The PAF precursors, lysoPAF and alkyl-acyl-glycerylphosphorylcholine (GPC), were also associated with the lamellar body fraction found in amniotic fluid. It has been concluded that the PAF in amniotic fluid associated with lamellar body surfactant is of fetal lung origin.

The possibility that PAF found in amniotic fluid originated in lung tissue was an attractive working hypothesis, not only from the standpoint that the fetal lung may participate in providing one of the signals involved in parturition but also because PAF may have a direct function in fetal lung maturation. A role of PAF in glycogenolysis has been proposed by Hanahan (for review, see Reference 61). It has been suggested that fetal lung glycogen can serve as a precursor of the surfactant glycerophospholipids.[62-64] The concentration of PAF and its lipid precursors was determined in lung and liver tissue of fetal rabbits throughout gestation.[65] The concentration of PAF in fetal rabbit lung increased some threefold between Day 21 and Day 31 of gestation whereas that in liver did not change. A similar increase in PAF concentration was found in human fetal lung tissue in organ culture.[65]

The activities of some of the enzymes involved in PAF metabolism were assayed in lung tissue throughout gestation. A Ca^{2+}-independent phospholipase A_2 with a substrate specificity similar to that described in amnionic tissue was also present in fetal rabbit lung.[66] The activity of lysoPAF:acetyl-coenzyme A (CoA) acetyltransferase increased threefold between the 21st and 24th day of gestation and remained elevated until birth in fetal rabbit lung[65] and human fetal lung explants placed in organ culture.[67] The activity of the specific choline phosphotransferase involved in the *de novo* pathway for PAF biosynthesis also increased.[68] Based on these observations, it was suggested that the combination of the enzymes involved in the remodeling as well as in the *de novo* pathway may account for the increase in PAF found in fetal lung during the latter stages of gestation. We found that the increase in PAF concentration, the decrease in glycogen content, and the increase in surfactant biosynthesis were temporally related.[35]

The role of PAF and its role in glycogen breakdown in fetal lung tissue was directly assessed.[69] Forty-five minutes after the IP injection of PAF to 24-day fetal rabbits *in utero,* fetal pulmonary and hepatic glycogen concentrations were significantly reduced. Lactate concentrations in fetal lung, liver, and plasma were increased. When either

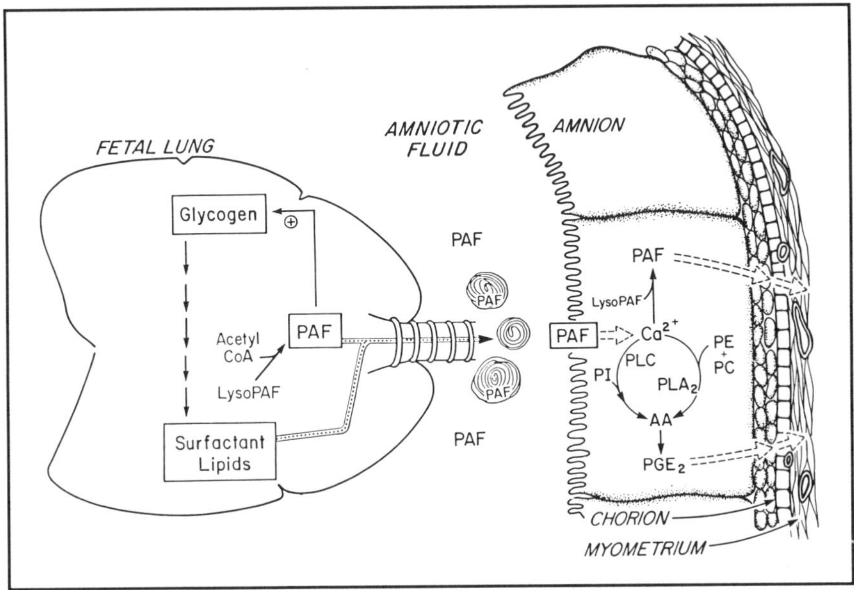

Figure 3. Role of fetal lung in regulation of PAF metabolism during parturition. Key: PI = phosphatidylinositol, AA = arachidonic acid, PE = phosphatidylethanolamine, PC = phosphatidylcholine, PLC = phospholipase C, PLA_2 = phospholipase A_2.

SRI-63-441 (antagonist of PAF) or the PAF antagonist plus PAF were injected, no effect on glycogen breakdown or lactate formation was found compared with the controls. Similarly, the injection of the enantiomer of the natural occurring *sn*-3 PAF, *sn*-1 PAF, did not significantly alter the glycogen or lactate content of fetal lung, liver, or plasma. The importance of PAF during lung development was further suggested by the observations of Kumar and Hanahan[70] and our group[71] that PAF markedly stimulates surfactant release by type II pneumonocytes.

The role of PAF in fetal lung maturation is summarized in Figure 3. It is suggested that the last major organ system to develop in the fetus, namely the lung, produces and secretes surfactant with its associated PAF into the amniotic fluid. The amniotic fluid PAF interacts with amnionic membranes to stimulate the production and release of PGE_2.[72] PAF may autocatalytically cause an increase in PAF synthesis and release, because PAF will increase intracellular Ca^{2+},[73] which in turn increases lysoPAF:acetylCo-A acetyltransferase activity in amnion,[74] as has been previously suggested for endometrium. A similar autocatalytic synthesis of PAF has been shown to occur in neutrophils.[75,76]

It has been suggested that the plasmalogen derivative of glycerophosphorylethanolamine (GPE) may also be involved in PAF and arachidonic acid metabolism. When a kidney cell line was preincubated with either alkyl-labeled lysoPAF or alkyl-acyl-GPC and subsequently challenged with various stimuli, it was shown that more than 30 percent of the radiolabel was recovered in the ethanolamine plasmalogen fraction.[77] Baker[78] has reported that PAF is primarily converted to the plasmalogen ethanolamine derivative in a macrophage cell line. Amnionic tissue obtained at term and in early labor contains a significant amount of the plasmalogen ethanolamine species (>75 percent of the total phosphatidylethanolamine fraction at this gestational age). The plasmalogen was also enriched with arachidonic acid.[79] When amnionic tissue was obtained between 12 and 17 weeks of gestation, almost equal amounts of the plasmalogen and diacyl species of ethanolamine glycerophospholipids were present. It is suggested that PAF is secreted by the fetal lung, in association with lamellar bodies, into the amniotic fluid and accounts for, in part, the presence of PAF in the amniotic fluid during early labor. As discussed, PAF in amniotic fluid stimulates the synthesis and release of PGE_2 from amnionic tissue as well as the possible further synthesis of PAF by this tissue by Ca^{2+} activation.[80] In amnionic tissue, therefore, one might expect an increase in the ethanolamine plasmalogen fraction due to PAF stimulation. Indeed, this was the case, because the ratio of plasmalogen to diacyl-GPE was increased from 1:1 in early membranes at 12 to 17 weeks to 3:1 at term (38 to 42 weeks). These findings provide further indirect support for the concept that PAF may directly affect amnionic tissue.

The inactivation of PAF is catalyzed by the enzyme PAF acetylhydrolase.[81] Two forms of the enzyme exist: one is found in the cytosolic fraction of most cells, and an extracellular form of the enzyme is present in plasma. The latter plasma enzyme was originally identified as the "acid labile factor."[82] Several reports have appeared concerning the change in activity of PAF acetylhydrolase in plasma in association with certain biological events. A significant increase in activity has been reported during a "stress" reaction in the lizard.[83] The activity also increased during platelet aggregation in the human[84] and in patients with ischemic cerebrovascular disease.[85] A decrease in PAF acetylhydrolase activity in maternal plasma was found during the second half of pregnancy in the rabbit[86] and the human.[35] The activity of this enzyme is increased dramatically in the plasma of the rabbit neonate shortly after delivery.[86] A decrease in plasma acetylhydrolase activity has also been reported in asthmatic children,[87] and a deficiency in serum PAF acetylhydrolase is transmitted as an autosomal recessive trait. An increase in PAF acetylhydrolase has been reported in the human in insulin-dependent diabetes mellitus[88] and also in spontaneously hypertensive rats[89] and hypertensive Caucasian human males.[90] The properties of PAF acetylhydrolase and its potential role in human disease have recently been reviewed in detail.[59]

In order to ascertain whether or not the capacity to inactivate PAF may be altered throughout gestation, the activity of PAF acetylhydrolase in maternal plasma was assayed in maternal and fetal rabbit plasma.[86] Inactivation of the PAF produced in the fetal compartment prior to its contact with the myometrium would be of considerable importance, because PAF has been shown to be a potent stimulator of myometrial contraction in a number of species.[52,91,92] The activity of PAF acetylhydrolase in the rabbits studied was approximately 130nmol x min^{-1} x ml^{-1} plasma prior to insemination.[86] During the first 15 days of pregnancy, the activity increased slightly and then dramatically decreased to 10nmol x min^{-1} x ml^{-1} plasma between the 23rd and 29th day of gestation. Following delivery, the activity returned to prepregnancy levels. The decrease in activity was not due to the presence of an inhibitor or activator. We also analyzed the activity in fetal and neonatal rabbit plasma starting on Day 21 of gestation.[86] In the 21st day fetal rabbit plasma, the activity was less than 20 percent of that found in maternal plasma. However, during the latter stages of gestation, the ratio reversed and the activity in fetal plasma was several times higher than that found in the corresponding maternal plasma, reaching values two to four times higher than that in nonpregnant maternal plasma during the postnatal period.[86] It is suggested that one of the reasons for the high resistance of neonate and young rabbits to PAF infusion[93] is the high circulating activity of PAF acetylhydrolase.

PAF Effect on Myometrium

Nishihira *et al.*[52] were the first to report that PAF propagated the contraction of strips of rat uterus. PAF also induces uterine contraction in the guinea pig[91] and the human.[92] The presence of PAF in myometrium has been established.[20,21] The presence of specific PAF receptors in the membrane fraction of both rabbit and human myometrium has recently been reported.[28,93]

Our group has also characterized the role of PAF in human myometrial smooth muscle cells in culture.[28] It was found that the addition of PAF at concentrations as low as 10^{-13}M to myometrial cells in culture resulted in a doubling of intracellular Ca^{2+} and a tripling of the phosphorylation of the myosin light chain.

The Effect of the Oral Administration of PAF Receptor Antagonists to Pregnant Rats

To further evaluate the involvement of PAF in the process of parturition, we administered two PAF receptor antagonists, SRI-63-441 and L-659,989, to 17-day timed pregnant rats and followed the events of labor and delivery. Animals given a high dose (IP, 20mg/kg/day) of SRI-63-441 showed first signs of delivery 12 hours before vehicle controls, but actual delivery of the litter was prevented in three of six animals by fetal blockage of the vaginal canal. A pronounced lengthening of the duration of parturition was found in rats treated with both PAF receptor antagonists. A 20-mg dose of SRI-63-441 resulted in a tenfold increase in the time required for parturition, although a high fetal mortality was evident. *Per os* administration of L-659,989 at two concentrations (2 and 20mg/kg/day) did not alter the gestational period; however, the duration of parturition was increased from two- to fourfold by such treatment. Again, no change in the length of gestation was noted with the PAF receptor antagonist L-659,989. Also, no side effects were observed at either concentration of the antagonist. Based on these experiments, it is suggested that PAF is an important mediator in the regulation of myometrial contraction necessary for expulsion of the fetus.

Hormonal Regulation of PAF Metabolism

Our group has recently investigated the mechanism(s) involved in the hormonal regulation of PAF acetylhydrolase.[94] 17α-Ethynylestradiol was administered to female and male rats, and the plasma PAF acetylhydrolase activity was found to decrease to one fifth in the vehicle-injected rats. A decrease was observed when an amount of estrogen as low as 50μg/kg was employed. In contrast, the injection of dexamethasone (IP, 1.3mg/kg body weight, five days) to male and female rats resulted in a threefold increase in the plasma PAF acetylhydrolase activity. The activity returned to the values prior to hormone-treatment four days after cessation of treatment. Testosterone and progesterone were without effect on the plasma PAF acetylhydrolase activity. The change in this activity caused by estrogen and the glucocorticoid was reflected by a change in the activity in the high-density lipoprotein fraction and not due to the presence of an inhibitor or activator in the plasma of the hormone-treated animals. Human plasma obtained from a group of women in which the 17β-estradiol concentration was elevated in preparation for an *in vitro* fertilization procedure showed an inverse relationship between plasma estrogen concentration and PAF acetylhydrolase activity. These findings suggested that estrogen is responsible for the regulation of PAF acetylhydrolase and the decrease in PAF acetylhydrolase activity during the latter stages of pregnancy in both the maternal and fetal plasma due to the hyperestrogenic state that occurs during this period.

The observed increase in PAF acetylhydrolase activity following dexamethasone treatment may account for, in part, the known antiinflammatory properties of this steroid concentration. This increase may also explain the reported increase in PAF acetylhydrolase activity found in the "stressed" lizard[83] and would be in agreement with previous observations on the alteration of PAF acetylhydrolase by hormonal treatment. Pritchard[95] demonstrated that the injection of rats with 17α-ethynylestradiol caused a decrease in both the lipoprotein fraction and the PAF acetylhydrolase activity in plasma.

PAF Metabolism in Complicated Pregnancies

Premature labor and premature rupture of membranes. Recently our group has determined the PAF concentration in the amniotic fluid of women whose pregnancies were destined to preterm delivery.[96] As a control, we analyzed the PAF present in amniotic fluid obtained from a group of patients with uncomplicated pregnancies at term, both in labor and not in labor. In confirmation of earlier observations, an eightfold increase in the PAF concentration of the PAF in the amniotic fluid obtained from "term-in-labor" samples compared with that of the "term-not-in-labor" samples was determined. The PAF concentration in the amniotic fluid of patients with "preterm labor" and "premature rupture of membranes" was also determined. The mean gestational age of both of these groups was approximately 32 weeks. There is a 20-fold elevation in the PAF concentration in the amniotic fluid of patients with "preterm labor" compared with the "term-not-in-labor" group. PAF concentration was also elevated to a similar extent (750fmol/ml) in the amniotic fluid obtained from a group of patients with "premature rupture of membranes." In the preterm labor group, two of the patients had microbial infections, and in the group with premature rupture of membranes, three of the patients also had infections. In the limited number of samples available, *no* differences in PAF concentration were found in the amniotic fluid obtained from the noninfected group comparedwith the infected group.[96]

Pregnancy-induced hypertension. The observation that PAF is a potent hypotensive agent[97] directed us to investigate the role of PAF in pregnancy-induced hypertension. We found that the PAF acetylhydrolase activity was approximately 44nmol x min^{-1} x ml^{-1} plasma in a group of nonpregnant women. As was the case in the rabbit, the activity of this enzyme in the human decreased during pregnancy and, by 32 weeks of gestation, was approximately 27nmol x min^{-1} x ml^{-1} in normal pregnant women. We found *no* decrease in PAF acetylhydrolase activity in seven patients with pregnancy-induced hypertension[35] and a mean gestational age of 32 weeks. It has been reported that patients with both mild and severe forms of this disease have decreased plasma estrogen levels. This observation would be in agreement with the reported effect of estrogen on PAF acetylhydrolase that was previously discussed.[94] We speculate that in patients with pregnancy-induced hypertension, the high PAF acetylhydrolase activity results in a decrease in the concentration of the hypotensive lipid PAF compared with the normotensive pregnant group. Thus, decreased capacity to modulate the various vasopressor agents may occur in patients with pregnancy-induced hypertension due to a decrease in the concentration of PAF. Future experiments are necessary in order to address the questions: "Is PAF the elusive compound that modifies the pressor response during pregnancy?" and "How is this potent second messenger regulated?"

In Figure 4 are summarized our group's findings concerning the role of PAF in relation to fetal lung development and the initiation of parturition. It is suggested that early in gestation the fetal tissues, including the fetal lung, have a limited capacity for PAF biosynthesis and that the small amount of PAF that reaches the amniotic fluid is inactivated by the PAF acetylhydrolase known to be present. Late in gestation, the fetal lung, and perhaps other fetal tissues, have an increased capacity for PAF biosynthesis. The increased PAF concentration in the type II pneumonocyte may facilitate the breakdown of glycogen, which serves, in part, as a carbon and energy source for the synthesis of the glycerophospholipids of surfactant and to stimulate surfactant secretion. During the latter stages of gestation, the lung secretes increased amounts of PAF in association with the surfactant-containing lamellar bodies that are secreted into the amniotic fluid. The increase in PAF synthesis ultimately exceeds the capacity of PAF

Figure 4. The proposed interrelations of PAF synthesized and secreted by fetal lung into amniotic fluid.

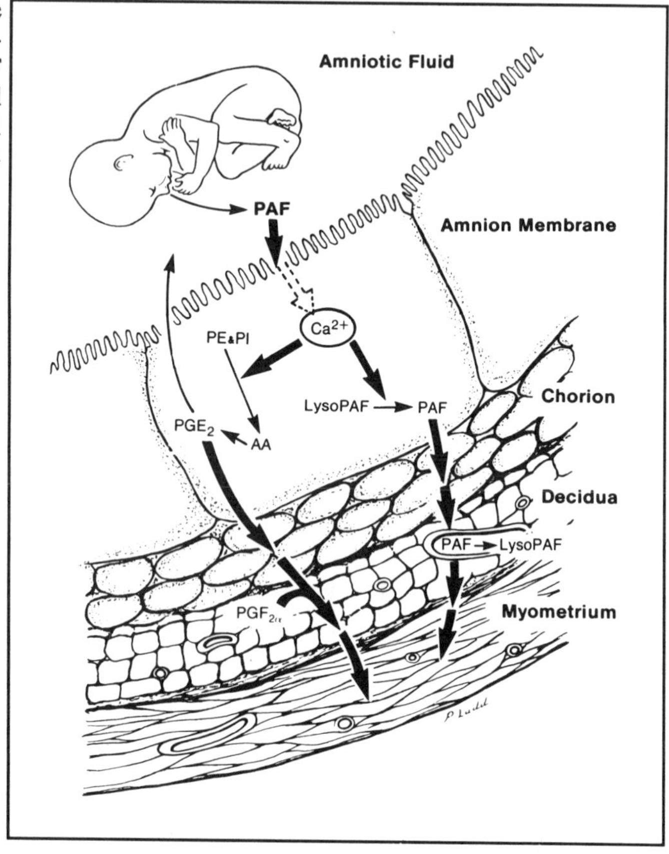

to be inactivated by the PAF acetylhydrolase[51] in amniotic fluid. The PAF in amniotic fluid interacts with fetal membranes to stimulate the release of arachidonic acid from diacyl-phosphatidylethanolamine and phosphatidylinositol catalyzed by phospholipases A_2 and C, possibly by increasing the Ca^{2+} concentration. Additional evidence in support of a role for PAF in activating amnionic tissue for autacoid formation is suggested by the observation that the ethanolamine plasmalogen fraction of human amnionic tissue increased during the latter stages of gestation.[98] It has been previously reported that ethanolamine plasmalogens are major metabolic end products of PAF in several tissues.[77,78] The increase in arachidonic acid results in an increase in substrate availability for prostaglandin biosynthesis that occurs during labor. In addition, the synthesis of PAF may also be stimulated in amnionic tissue because it is known that small increases in Ca^{2+} markedly increase PAF biosynthesis. It is now well established that PAF stimulates PAF synthesis in a number of cell types.[99,100] Early in pregnancy, the small amounts of PAF that may reach the myometrium may be inactivated by the high circulating levels of PAF acetylhydrolase known to be present in the maternal plasma. The "lush" blood supply of the decidua may be the site of inactivation of PAF. However, during the latter stages of pregnancy, when an increase in PAF synthesis is occurring in fetal lung and other fetal tissues, there is a *decrease* in the capacity of the maternal plasma to inactivate PAF due to the decrease in acetylhydrolase activity. Therefore, the PAF, along with prostaglandins, may now reach the myometrium, where both compounds are known to stimulate contractions. Thus, it would appear that the last major organ system to mature, namely the fetal lung, *may participate* in the *signaling* from the fetus to the myometrium during the events of parturition.

Acknowledgments

JMJ gratefully acknowledges the research contributions of his colleagues, Dr. T. Okazaki, G.C. Di Renzo, N. Sagawa, M.M. Billah, M.M. Anceschi, C. Ban, D.R. Hoffman, M. J. Angle, S. Miyaura, Y.-P. Zhu and J.E. Bleasdale. Several of these colleagues were the recipients of fellowships from the Chilton Foundation in Dallas, Texas. We also gratefully acknowledge the editorial assistance of Ms. Dolly Tutton. Investigations performed in the author's laboratories were supported by U.S. Public Health Service Grants HD13912 and HD11149 and the Robert A. Welch Foundation, Houston, Texas.

References

1. P. Braquet *et al., Pharmacol Rev.* **39**, 97 (1987).
2. C.A. Demopoulos, R.N. Pinckard, and D.J. Hanahan, *J. Biol. Chem.* **254**, 9355 (1979).

3. M.L. Blank *et al., Biochem. Biophys. Res. Commun.* **90**, 1194 (1979).
4. J. Benveniste *et al., C. Rend. Acad. Sci.* **389D**, 1037 (1979).
5. A.O. Abisogun, P. Braquet, and A. Tsafriri, *Science* **243**, 381 (1989).
6. L.L. Espey *et al., Biol. Reprod.*, in press.
7. C. Orozco, T. Perkins, and F.M. Clarke, *J. Reprod. Fertil.* **78**, 549 (1986).
8. K. Sueoka *et al., Am. J Obstet. Gynecol.* **159**, 1580 (1988).
9. R. Kumar, M.J.K. Harper, and D.J. Hanahan, *Arch. Biochem. Biophys.* **260**, 497 (1988).
10. B.S. Minhas *et al., Fertil. Steril.*, 44th Annual Meeting, Program Suppl. **522**, 23 (Abstr. 065) (1988).
11. D.D. Ricker *et al., Theriogenology* **31**, 247 (Abstr.) (1989).
12. C. O'Neill, *J. Reprod. Fertil.* **73**, 566 (1985).
13. C. O'Neill *et al., Ann. NY Acad. Sci.* 442, 429 (1985).
14. C. O'Neill, *Platelet-Activating Factor and Human Disease,* P. Barnes, C.P. Page, and P.M. Henson, Eds. (Blackwell Scientific Publications, Oxford,England, 1990), pp. 282-296.
15. C. O'Neill, *J. Reprod. Fertil.* **75**, 375 (1985).
16. C. O'Neill *et al., Fertil. Steril.* **47**, 969 (1987).
17. M.L. Amiel *et al., Hum. Reprod.* **4**, 327 (1989).
18. C. O'Neill *et al., Lancet* **30**, 769 (1989).
19. M.J. Angle and J.M. Johnston, *Uterine Function Molecular and Cellular Aspects,* M. Carsten and J.D. Miller, Eds. (Plenum Press, New York, 1990), in press.
20. K. Yasuda, K. Satouchi, and K. Saito, *Biochem. Biophys. Res. Commun.* **138**, 1231 (1986).
21. M.J. Angle *et al., J. Reprod. Fertil.* **83**, 711 (1988).
22. A.A. Alecozay *et al., Biol. Reprod.*, in press.
23. M. Nonogaki *et al.*, Third International Conference on Platelet Activating Factor and Structurally Related Alkyl Ether Lipids, Abstract #L-44, 1989, p. 44.
24. A.A. Alecozay *et al., The Endocrine Society 71st Annual Meeting,* Program and Abstracts, in press.
25. S.K. Smith and R.W. Kelly, *J. Reprod. Fertil.* **82**, 271 (1988).
26. M.M.Billah *et al., Prostaglandins* **30**, 841 (1985).
27. G.B. Kudolo and M.J.K. Harper, *Biol. Reprod.*, in press.
28. Y.-P. Zhu, R.A. Word, and J.M. Johnston, *Proceedings of the Society for Gynecologic Investigations,* 37th Annual Meeting, No. 130, p. 161 (1990).
29. I.F. Lau, S.K. Saksena, and M.C. Chang, *Prostaglandins* **4**, 795 (1973).
30. S.K. Saksena, I.F. Lau, and M.C. Chang, *Acta Endocrinol. (Copenh.)* **81**, 801 (1976).
31. T.G. Kennedy, *Biol. Reprod.* **16**, 286 (1977).
32. C.A. Phillips and N.L. Poyser, *J. Reprod. Fertil.* **62**, 73 (1981).
33. G.D. Thorburn, *Initiation of Parturition: Prevention of Prematurity* (Ross Laboratories, Columbus, OH, 1983), pp. 2-8.
34. P.C. MacDonald *et al., Obstet. Gynecol.* **44**, 629 (1974).
35. J.M. Johnston, *Platelet-Activating Factor and Diseases* (International Medical Publishers, Tokyo, Japan, 1989), pp. 129-151.
36. T.U. Luukkainen and A.I. Csapo, *Fertil. Steril.* **14**, 629 (1963).
37. J.T. Lanman, L. Herod, and R. Thau, *Pediatr. Res.* **6**, 701 (1972).
38. J.T. Lanman, L. Herod, and R. Thau, *Peditar. Res.* **8**, 1 (1974).
39. Y. Ogawa, L. Herod, and J.T. Lanman, *Gynecol. Invest.* **1**, 240 (1970).

40. P.W. Nathanielsz, M. Abel, and G.W. Smith, *Fetal and Neonatal Physiology* (Cambridge University Press, Cambridge, NY, 1973), pp. 594-601.
41. F. Hertelendy, M. Yeh, and H.V. Bielier, *Gen. Comp. Endocrinol.* **22**, 529 (1973).
42. W.E.M. Lands and B. Samuelsson, *Biochim. Biophys. Acta* **164**, 429 (1958).
43. H. Vonkeman and D.A. Van Dorp, *Biochim. Biophys. Acta* **164**, 426 (1958).
44. S.M.M. Karim, *J. Obstet. Gynecol. Br. Commonw.* **73**, 903 (1966).
45. J.R. Okita, P.C. MacDonald, and J.M. Johnston, *J. Biol. Chem.* **257**, 14029 (1982).
46. J.E. Bleasdale and J.M. Johnston, *Reviews in Perinatal Medicine, Vol. 5* (Alan R. Liss, NY, 1984), pp. 151-191.
46. G.C. Di Renzo *et al., J. Clin. Invest.* **67**, 847 (1981).
47. T. Okazaki *et al., Biol. Reprod.* **25**, 103 (1981).
48. P.M. Majerus *et al., Advances in Prostaglandin, Thromboxane, and Leukotriene Research, Vol. 7* (Raven Press, NY, 1983), pp. 45-52.
49. N. Sagawa *et al., J. Biol. Chem.* **257**, 8158 (1982).
50. F. Snyder, *Med. Res. Rev.* **5**, 107 (1985).
51. M.M. Billah and J.M. Johnston, *Biochem. Biophys. Res. Commun.* **113**, 51 (1983).
52. J. Nishihira *et al., Lipids* **19**, 907 (1984).
53. M.M. Billah *et al., Prostaglandins* **30**, 841 (1985).
54. J.E. Bleasdale *et al., Proceedings of IX European Congress of Perinatal Medicine* (Harwood Academic Publishers, Oxford, England, 1989), pp. 828-834.
55. E. Ninio *et al., Biochim. Biophys. Acta* **710**, 23 (1982).
56. D.J. Lenihan and T.-C. Lee, *Biochem. Biophys. Res. Commun.* **120**, 834 (1984).
57. T. Okazaki, C. Ban, and J.M. Johston, *Arch. Biochem. Biophys.* **299**, 27 (1984).
58. J. Gomez-Cambronero *et al., Biochem. J.* **237**, 439 (1986).
59. M.R. Elstad *et al., Platelet-Activating Factor and Diseases* (International Medical Publishers, Tokyo, Japan, 1989), pp. 69-84.
60. R. Kumar, R.J. King, and D.J. Hanahan, *Biochim. Biophys. Acta* **836**, 19 (1985).
61. D.J. Hanahan, *Ann. Rev. Biochem.* **55**, 483 (1986).
62. J.R. Bourbon *et al., Biochim. Biophys. Acta* **712**, 382 (1982).
63. P.M. Farrell and J.R. Bourbon, *Biochim. Biophys. Acta* **878**, 159 (1986).
64. W.M. Maniscalco *et al., Biochim. Biophys. Acta* **530**, 333 (1978).
65. D.R. Hoffman, T.C. Truong, and J.M. Johnston, *Biochim. Biophys. Acta* **879**, 88 (1986).
66. M.J. Angle, F. Paltauf, and J.M. Johnston, *Biochim. Biophys. Acta* **962**, 234 (1989).
67. D.R. Hoffman, C.T. Truong, and J.M. Johnston, *Ann. J. Obstet. Gynecol.* **155**, 70 (1986).
68. D.R. Hoffman, M. Bateman, and J.M. Johnston, *Lipids* **23**, 96 (1988).
69. D.R. Hoffman *et al., J. Biol. Chem.* **263**, 9316 (1989).
70. R. Kumar and D.J. Hanahan, *Platelet-Activating Factor and Related Lipid Mediators* (Plenum Publishing Corp., New York, 1987), pp. 239-254.
71. J.M. Johnston, J.E. Bleasdale, and D.R. Hoffman, ibid., pp. 375-402.
72. T. Okazaki *et al., Biol. Reprod.* **25**, 103 (1981).
73. F. Snyder, *Med. Res. Rev.* **5**, 107 (1985).
74. M.M. Billah *et al., Prostaglandins* **30**, 841 (1985).
75. T.W. Doebber and M.S. Wu, *Proc. Natl. Acad. Sci. USA* **84**, 7557 (1987).
76. T.G. Tessner, J.T. O'Flaberty, and R.L. Wykle, *Fetal Medicine Review* (Blackwell Scientific Publications, Oxford, England, 1989), pp. 4794-4799.
77. LW. Daniel, M. Waite, and R.L. Wykle, *J. Biol. Chem.* **261**, 9128 (1986).

78. R. Baker, *FASEB J.* **3**, A1377 (Abstr.) (1988).
79. J.R. Okita, Ph.D. Dissertation (The University of Texas Health Science Center at Dallas, 1981).
80. C. Ban *et al., Arch. Biochem. Biophys.* **246**, 9 (1986).
81. T.-C. Lee and F. Snyder, *Platelet-Activating Factor and Human Diseases* (Blackwell Scientific Publishers, Oxford, England, 1989), pp. 1-22.
82. R.S. Farr *et al., Clin. Immunol. Immunopathol.* **15**, 318 (1980).
83. D.J. Lenihan, N. Greenberg, and T.-C. Lee, *Comp. Biochem. Physiol.* **81C**, 81 (1985).
84. Y. Suzuki *et al., Eur. J. Biochem.* **712**, 117 (1988).
85. K. Satoh *et al., Prostaglandins* **35**, 685 (1988).
86. N. Maki, D.R. Hoffman, and J.M. Johnston, *Proc. Natl. Acad Sci. USA* **85**, 728 (1988).
87. M. Miwa *et al., J. Clin. Invest.* **82**, 1983 (1988).
88. B. Hofmann *et al., Haemostasis* **19**, 180 (1989).
89. M.L. Blank *et al., Biochem. Biophys. Res. Commun.* **113**, 666 (1983).
90. J.E. Crook *et al., Clin. Sci.* **74**, 393 (1988).
91. G. Montrucchio *et al., Prostaglandins* **32**, 539 (1986).
92. G. Tetta *et al., Proc. Soc. Exp. Biol. Med.* **183**, 376 (1986).
93. M.J.K. Harper, *Biol. Reprod.* **40**, 907 (1989).
94. S. Miyaura *et al., Lipids,* in press.
95. P.H. Pritchard, *Biochem. J.* **246**, 791 (1987).
96. D.R. Hoffman, R. Romero, and J.M. Johnston, *Ann. J. Obstet. Gynecol.* **162**, 525 (1990).
97. F. Snyder, *Proc. Soc. Exp. Biol. Med.* **190**, 125 (1989).
98. J.M. Johnston *et al., Eicosanoids in Reproduction* (CRC Press, Boca Raton, FL, 1990), in press.
99. T.W. Doebber and M.S. Wu, *Proc. Natl. Acad. Sci. USA* **84**, 7557 (1987).
100. T.G. Tessner, J.T. O'Flaherty, and R.L. Wykle, *J. Biol. Chem.* **264**, 4794 (1989).

9

Modeling of PAF Pathophysiology:

Lethal Shock Induced by PAF in Rodents

Adam K. Myers
Peter W. Ramwell
Department of Physiology and Biophysics
Georgetown University Medical Center
Washington, DC 20007

Platelet activating factor, an important mediator of sepsis and anaphylaxis, produces circulatory shock and dose-dependent death when administered intravenously. PAF-induced mortality in mice has proven to be an interesting and useful model for several purposes. Pharmacologically, PAF toxicity in mice is used for *in vivo* screening of PAF antagonists. Physiologically, PAF-induced death is interesting from the standpoint of compensatory mechanisms in shock. Both an intact sympatho-adrenomedullary system and a functional pituitary-adrenocortical axis are apparently necessary for adaptive, compensatory responses to the cardiovascular derangements associated with PAF toxicity. Studies of PAF lethality have also been valuable in the investigation of mediators of sepsis and endotoxemia. Comparison of mouse mortality models involving injection of PAF, endotoxin, or tumor necrosis factor (TNF) demonstrate that endotoxin toxicity in mice is mediated by PAF, among other factors. Studies in these models also suggest that TNF has limited toxicity by itself, but probably acts synergistically with other mediators of endotoxemia in producing the pathophysiology associated with sepsis.

Introduction

The importance of platelet activating factor (PAF) in the cardiorespiratory sequelae of sepsis and anaphylaxis is now apparent. PAF is released when animals are challenged by anaphylactic stimuli[1] or bacterial endotoxin.[2] Sources of PAF include leukocytes and related cells, platelets, and vascular endothelium.[3] The effects of challenge by exogenous PAF are consistent with a central role in the pathophysiological responses to sepsis or anaphylactic stimuli. PAF is a potent bronchoconstrictor, and in the cardiovascular system its administration induces hypotension, cardiac depression, hemoconcentration, and species-dependent platelet activation.[4-7] Finally, pharmacological antagonists of PAF are protective in a number of models of anaphylaxis and sepsis.[2,8,9]

Owing to this central role of PAF in some forms of shock, *in vivo* models of the systemic pathophysiology of PAF are of substantial interest. One model which has proved useful in studying pharmacological measures for blocking deleterious systemic effects of PAF is PAF-induced death in mice.

Mechanism of PAF Lethality

Intravenous injection of PAF provokes dose-dependent death in the micromolar range in mice (Figure 1), within a few minutes to approximately an hour after PAF administration.[10,11] The mechanism of this death is apparently cardiovascular in nature, involving circulatory shock, hemoconcentration, and perhaps bronchoconstriction. Evidence for this conclusion includes the potent hypotensive and hemoconcentrating effects of PAF, as discussed in the preceeding text. Other evidence involves the spectrum of drugs which protect against or exacerbate PAF lethality in mice; these drugs will be discussed in more detail below. The role of thrombosis in PAF lethality is highly species dependent. In the rabbit, which is extremely sensitive to PAF platelet aggregatory actions, PAF injection provokes thrombotic death.[12] Murine platelets are insensitive to PAF[11,13] and

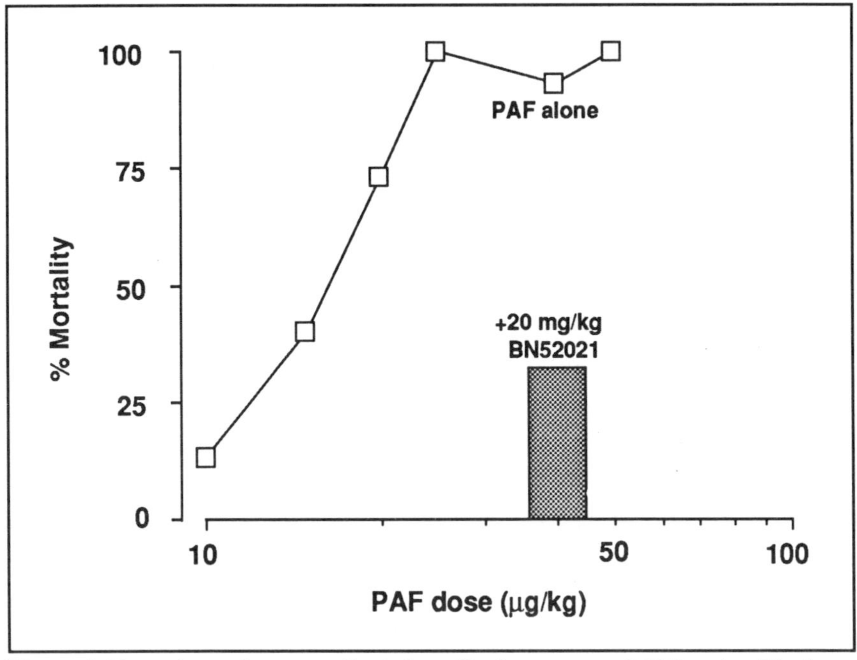

Figure 1. Dose-dependent mortality induced by intravenous PAF in mice. The inset bar illustrates the protective action of PAF antagonist pretreatment, in this case by BN 52021. This figure is derived from data in References 10 and 14.

thus thrombosis is not a factor in mice. In humans, platelets have intermediate sensitivity to PAF[13] and thus PAF might partially account for the platelet activation associated with shock states.

Effects of PAF Antagonists in the Mouse Model

As a class, PAF antagonists are potent and highly efficacious blockers of PAF toxicity in mice. Among the many drugs which have been tested and found protective are BN 52021,[14] L 652,731,[14] CV-3988,[9] CV-6209,[15] SDZ 64-412,[16] and kadsurenone.[17] Notably, PAF antagonists are not able to inhibit lethality induced by thromboxane agonists, whereas thromboxane antagonists that protect against the latter challenge are ineffectual against PAF.[14] Similarly, leukotriene antagonists such as FPL 55712 are also unable to block PAF induced death in our studies[14] and those of others[17] (although contradictory results have been reported[9,18]). These findings demonstrate the specificity of these sudden death models. The PAF lethality technique in mice has been suggested as an *in vivo* test for bioactivity of PAF antagonists,[14] and should prove useful for examining the bioavailability of the PAF antagonists by various routes of administration.

Adrenergic Modulation of PAF Toxicity

Beta adrenergic blockers are one of the few classes of drugs shown to potentiate PAF lethality in mice, and suggest the importance of sympatho-adrenomedullary mechanisms in compensating for the systemic effects of PAF. Several studies have shown that beta blockade prolongs PAF-induced hypotension in rats,[19,20] and plasma catecholamines are elevated greatly after PAF challenge.[21] Adrenal demedullation[19] has effects similar to beta blocker administration. In the mouse model, propranolol pretreatment enhances lethality of PAF while alpha adrenergic antagonists have little effect,[20] which we have interpreted as reflecting the involvement of hypotension and cardiac depression in the death. On the other hand, drugs which augment or directly stimulate sympathetic outflow by central mechanisms, for example, thyrotropin-releasing hormone and naloxone, are protective in this model.[20,22] The observation that positive inotropic drugs are beneficial[23] also supports a role for circulatory shock in the death.

In contrast to our interpretation, others have suggested that beta blockers enhance PAF toxicity by inhibiting adrenergic compensation for the bronchoconstrictive effects of PAF.[24] This view is based on the differential effects of beta-1 and beta-2 antagonists. However, because PAF does not promote platelet activation in mice or rats,[13] and hence does not cause the degree of release of the potent bronchoconstrictor thromboxane seen in some other species, bronchoconstriction is not a major effect of PAF in the rat and perhaps the mouse. This aspect of PAF toxicity thus remains unresolved; it is likely that both cardiovascular collapse and bronchoconstriction have some role in PAF pathophysiology in the mouse.

Corticosteroid Modulation of PAF Toxicity

One of our early observations in this area of study was the highly efficacious protection afforded by glucocorticoid pretreatment against PAF induced death.[10] In addition to affording nearly complete protection against the death, steroids can totally block the hemoconcentration associated with intravenous PAF in mice.[25] The beneficial action of steroids has been confirmed by others,[17,18] but we can still only speculate on its mechanism. Among the many, unproven possibilities are augmentation of beta adrenergic mechanisms (cardiovascular or respiratory) by steroids, since they are known to upregulate beta adrenergic mechanisms,[26] and inhibition of the release of various mediators, since steroids block release of many substances which participate in immune and inflammatory respones.[27]

Hypothetical Role of Adrenal in the Physiological Response to PAF

Based on these observations concerning adrenergic drugs and corticosteroids, we hypothesize that both sympathoadrenal and adrenocortical mechanisms are important physiological mechanisms in the compensation for systemic sequelae of PAF overflow in pathologic states. Of course, sympathetic and adrenocortical roles in responses to stress and shock are well-known, but it is interesting to reconsider these roles with respect to a specific mediator of shock such as PAF. In our hypothetical scheme illustrated in Figure 2, during septic or anaphylactic shock, PAF is released by leukocytes, platelets, endothelium and other tissues, and can provoke circulatory shock. This hypodynamic circulatory condition promotes reflexive sympathetic activation; catecholamines released by adrenergic nerve endings and adrenal medulla stimulate the cardiovascular system and dilate airways, resulting in compensation. Similarly, ACTH is released by the anterior pituitary; the subsequent release of corticosteroids by the adrenal cortex also promotes cardiorespiratory compensation as well as inhibits release of various mediators. It is noteworthy that the role

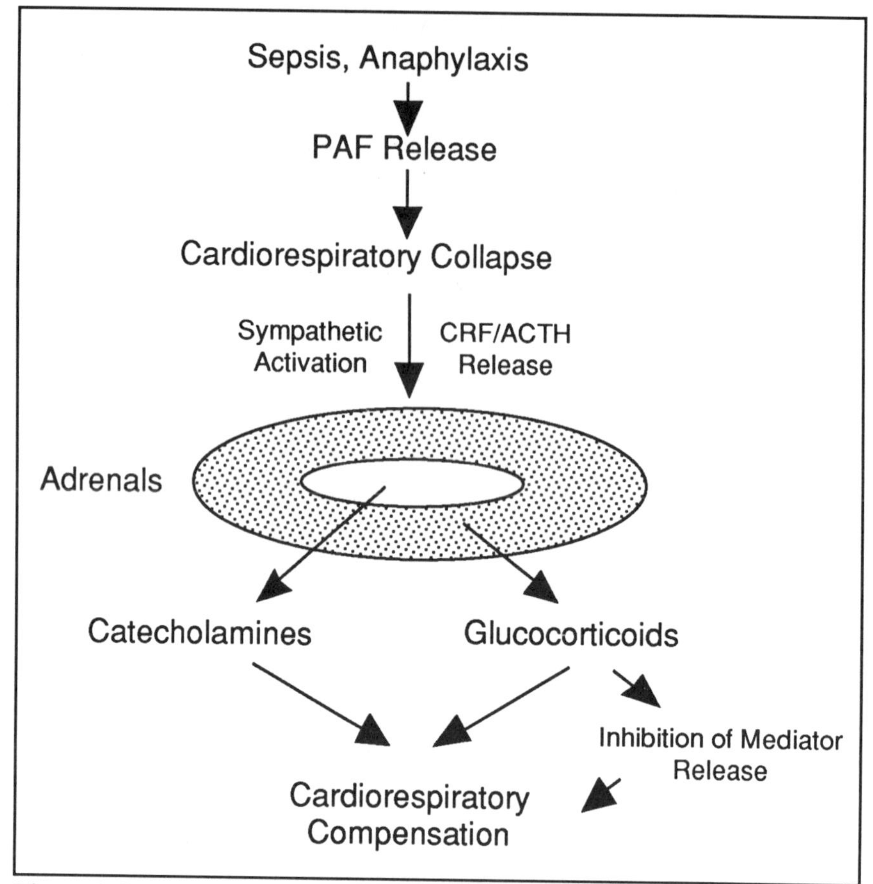

Figure 2. Proposed role of the adrenal gland in compensatory responses to PAF-induced circulatory shock in sepsis and anaphylaxis. The cardiovascular alterations provoked by PAF stimulate reflexive sympathetic activation as well as activation of the pituitary adrenal axis. Subsequently, adrenomedullary release of catecholamines and cortical release of glucocorticoids act as compensatory mechanisms. However, the role of the adrenal cortex is probably more critical in terms of the preexisting susceptibility to challenge rather than as an acute compensatory mechanism in lethal shock models, because steroids are most effective when used to pretreat animals.

of the adrenal cortex is probably less important in terms of compensation as opposed to preexisting resistance to perturbation. Typically, in these types of models, adrenalectomy greatly exacerbates susceptibility to challenge and pretreatment with glucocorticoids is highly beneficial, but post-treatment is not effective. Presumably, in a slowly evolving shock state both the basal level of glucocorticoid and the compensatory release of steroids would be important, while if the onset of shock is sudden, as it is in our model, there is insufficient time to invoke the action of newly synthesized glucocorticoids.

Relationship of PAF Toxicity to Endotoxin Toxicity

Another useful application of PAF toxicity in mice has been its comparison to endotoxin-induced death in mice. This technique has been used to demonstrate simultaneously the action of drugs such as SDZ 64-412 in specifically antagonizing PAF effects and protecting against endotoxin.[16] The efficacy of PAF antagonists in the two models has been cited as evidence that the drugs are active *in vivo* as PAF antagonists, and can protect against lethality in a closely related model of sepsis. We have also tested PAF antagonists against endotoxin lethality in mice and find that BN 52021 affords a high degree of protection against both the death and the hemoconcentration associated with endotoxin administration.[28] In these studies, high dose intravenous endotoxin results in rapid deterioration and death within a few hours, and this toxicity is blocked by pretreatment with the PAF antagonist, the cyclooxygenase inhibitor indomethacin or the dual cyclooxygenase-lipooxygenase inhibitor BW 755C.[28] Thus, high dose endotoxin-induced death in mice is mediated by both PAF and eicosanoids. Thromboxane A_2 and leukotrienes are probably the major eicosanoids involved, based on a large body of literature.

Relationships Between Tumor Necrosis Factor, PAF, and Endotoxin Lethality

In conjunction with studies on PAF and endotoxin toxicity, we have also examined the role of tumor necrosis factor (TNF). TNF, which is a major product of activated macrophages, has been proposed as the primary mediator of endotoxin-induced shock, capable of producing all of the major effects of endotoxin *in vivo*.[29] Further, passive immunization against TNF blocks the lethality of endotoxin.[30] However, recent experiments bring into question the relative importance of various mediators in endotoxin shock in mice. Studies in our laboratory,[28] as well as those of others[31] demonstrate that TNF is not highly toxic in mice, unless animals are primed with sublethal doses of endotoxin. Thus, our current view is that TNF is an important mediator of endotoxemia in mice, but it acts synergistically with other mediators, perhaps, for example, interleukin-1, a relationship which has been established in several systems including TNF lethality in mice.[32]

In addition, we have failed to observe any protective action of PAF antagonists or arachidonic acid metabolism inhibitors on the toxicity of TNF in combination with low-dose endotoxin.[28] One possible explanation is that the toxicity of the combination challenge (because it is slow in developing, requiring a few hours at least) is not readily blocked by pretreatment with the drugs. Regardless, our results strongly suggest that whereas high-dose endotoxin administration is mediated by the early release of lethal quantities of lipid mediators, TNF/low-dose endotoxin toxicity is less dependent on such an early release of PAF and eicosanoids, and TNF by itself is less toxic in mice than might be assumed from earlier studies.

Overall, based on our studies of endotoxin, PAF, TNF, and on the larger literature, we suggest the following hypothetical mediator sequence (Figure 3): Endotoxin challenge stimulates both the release of TNF and other cytokines, as well as PAF and other lipid mediators. The cytokines synergize to induce further release of PAF and eicosanoids. Ultimately, the lipid mediators are responsible in large part for the cardiorespiratory and other major derangements associated with endotoxemia. Of course, a variety of other classes of mediators are involved, as well as the direct effects of bacterial products.

Conclusions

Based on the foregoing, the mouse model of PAF-induced death is a useful tool for modeling the pharmacology and pathophysiology

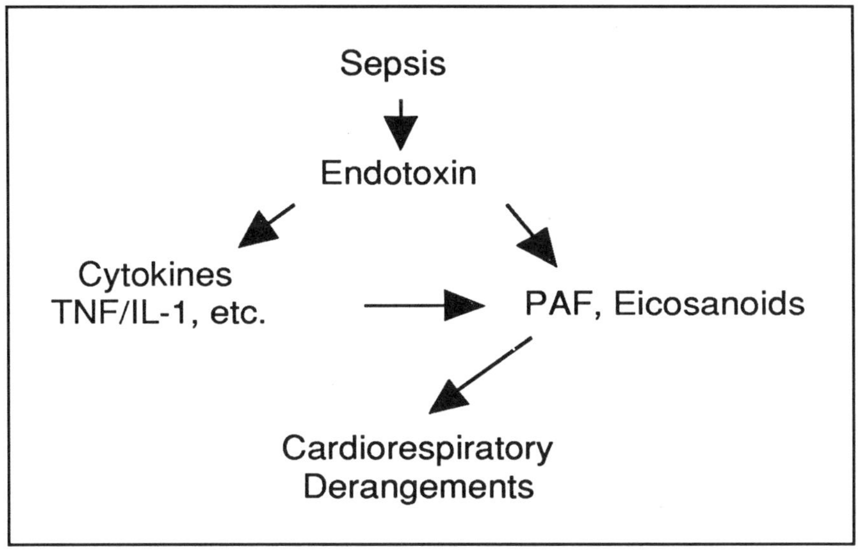

Figure 3. Hypothetical mediator sequence in models of sepsis such as endotoxin-induced mortality in mice. PAF and eicosanoids, along with other factors, are important mediators of cardiovascular derangements and lethality in these models. Contrary to early reports suggesting that TNF is the central mediator of endotoxic shock, capable of producing alone all of the major effects of endotoxin, more recent work suggests that TNF acts synergistically with other cytokines. In our scheme, endotoxin can directly stimulate release of lipid mediators such as PAF and eicosanoids, or indirectly, stimulate their release through the actions of TNF, interleukin-1, and other cytokines, acting in synergy. The cardiorespiratory collapse associated with endotoxemia is the result of complex interactions of many mediators, including those depicted here.

of PAF. It is an ideal model for testing *in vivo* activity of PAF antagonists. Furthermore, PAF toxicity can be used in studying compensatory systems in shock. Finally, in conjunction with other related models such as endotoxin-induced death, insights are developing into the interrelationships between various mediators of shock.

References

1. R.N. Pinckard *et al., J. Immunol.* **119**, 2185 (1977).
2. T.W. Doebber *et al., Biochem. Biophys. Res. Comm.* **1127**, 799 (1985).
3. P. Braquet *et al., Pharmacol. Rev.* **39**, 97 (1987).
4. L.M. McManus *et al., J. Immunol.* **124**, 2919 (1980).

5. M. Chignard *et al., J. Pharmacol. (Paris)* **11**, 371 (1980).
6. B.B. Vargaftig *et al., Eur. J. Pharmacol.* **65**, 185 (1980).
7. L.M. McManus *et al., Lab. Invest.* **45**, 303 (1981).
8. Z.-I. Terashita *et al., Eur. J. Pharmacol.* **109**, 257, (1985).
9. Z.-I. Terashita *et al., J. Pharmacol. Exp. Ther.* **243**, 378 (1987).
10. A. Myers, E. Ramey, and P. Ramwell, *Br. J. Pharmacol.* **79**, 595 (1983).
11. E. Lanara *et al., Biochem. Biophys. Res. Comm.* **109**, 1148 (1982).
12. A.M. Lefer, H.F. Muller, and J.B. Smith, *Br. J. Pharmacol.* **83**, 125 (1984).
13. D.H. Namm, A.S. Tadepalli, and J.A. High, *Thromb. Res.* **25**, 341 (1982).
14. A.K. Myers, T. Nakanishi, and P. Ramwell, *Prostaglandins* **35**, 447 (1988).
15. Z.-I. Terashita *et al., J. Pharmacol. Exp. Ther.* **242**, 263 (1987).
16. D.A. Handley *et al., J. Pharmacol. Exp. Ther.* **274**, 617 (1988).
17. R.P. Carlson, L. O'Neill-Davis, and J. Chang, *Agents Actions* **21**, 379 (1987).
18. J. M. Young *et al., Prostaglandins* **30**, 545 (1985).
19. Z. Zukowska-Grojec *et al., Clin. Exp. Hypertens.* **A7**, 1015 (1985).
20. A.K. Myers and A.P. Torres Duarte, *Int. J. Immunopath. Pharmacol.* **2**, 181 (1989).
21. G. Feuerstein *et al., Clin. Exp. Hypertens.* **A4**, 1335 (1982).
22. A. Myers, A.P. Duarte, and P. Ramwell, in *Advances in Prostaglandin, Thromboxane, and Leukotriene Research, Vol. 17*, B. Samuelsson, R. Paoletti, and P.W. Ramwell, Eds. (Raven Press, New York, 1987), pp 833-837.
23. D. Kelefiotis *et al., Life Sci.* **42**, 623 (1988).
24. M. Criscuoli and A. Subissi, *Br. J. Pharmacol.* **90**, 203 (1987).
25. A.K. Myers and T.J. Bader, *Circ. Shock* **23**, 143 (1987).
26. E. Boecklen, S. Flad, and H.V. Faber, *Drug Dev. Res.* **10**, 11, 1987.
27. A. Munck, P.M. Guyre, and N.J. Holbrook, *Endocrine Rev.* **5**, 25 (1984).
28. A.K. Myers, J.W. Robey, and R.M. Price, *Br. J. Pharmacol.* **99**, 499 (1990).
29. K.J. Tracey *et al., Science* **234**, 470 (1986).
30. B.A. Beutler, I.W. Milsark, and A.C. Cerami, *Science* **229**, 869 (1985).
31. J.L. Rothstein and H. Schreiber, *Proc. Natl. Acad. Sci. USA* **85**, 607 (1988).
32. A. Waage and T. Espevik, *J. Exp. Med.* **167**, 1987 (1988).

10

PAF Priming of Inflammatory Responses

Many potentially autotoxic agents are involved in inflammation. These pro-inflammatory mediators activate diverse cell types following binding to their respective membrane receptors through similar, often overlapping intracellular signal transduction pathways. Studies performed *in vitro* with isolated cell preparations demonstrate that low concentrations of mediators can interact synergistically to induce enhanced cellular responses. For example, substimulatory concentrations of platelet-activating factor (PAF) "prime" inflammatory effector cells for augmented responses to subsequent stimulation with other agonists and conversely, PAF-induced bioactions are primed by other mediators. Furthermore, in experimental animals, administration of small doses of mediators, which by themselves are innocuous, may induce severe inflammatory reactions when given in combination. Thus, priming can occur *in vivo*. Studies performed in our laboratories have shown that endotoxin primes isolated perfused rabbit lungs for a markedly enhanced response to PAF. These findings support the hypothesis that multiple pro-inflammatory mediators interact synergistically to augment injury. Priming and primed stimulation may be important in the pathogenesis of overwhelming inflammation that occurs in the setting of sepsis and the adult respiratory distress syndrome.

correspndence

William L. Salzer

Charles E. McCall

Section of Infectious Diseases
Department of Medicine
Bowman Gray School of Medicine of Wake Forest University
300 South Hawthorne Road
Winston-Salem, NC 27103

Introduction

Cellular stimulation produced by platelet-activating factor (PAF) occurs following binding to specific membrane receptors. This receptor-ligand complex activates pathways generating second messengers that operate intracellularly to translate the stimulus (PAF, in this instance) into cellular responses. In addition, cells that produce PAF do so following specific stimulation via these same stimulus-response coupling mechanisms. Evidence also suggests that PAF itself may function intracellularly as a second messenger or as an autocrine agent because most cell types that produce PAF are stimulated by PAF.

Studies *in vitro* and *in vivo* have defined the numerous target cells and physiological responses induced by PAF. However, these effects are often elicited by PAF concentrations far in excess of those that might occur locally or systemically with endogenous PAF production in the whole organism. This raises questions as to how small amounts of endogenous PAF could operate *in vivo*. One possible explanation for *in vivo* responsiveness to seemingly small amounts of a mediator could be synergistic interactions between multiple mediators. In the course of local or systemic inflammatory responses, a vast array of potential pro-inflammatory agents are produced. These may recruit and activate other inflammatory effector cells and thereby amplify ongoing reactions. In this milieu, the effects of a single mediator might be modulated by the effects of other mediators.

Evidence for such synergistic interactions between agonists has been obtained *both in vitro* and *in vivo*. Studies with polymorphonuclear leukocytes (PMN) have shown that exposure to substimulatory concentrations of an agonist induces a "primed" state such that exposure to a second agonist elicits markedly enhanced responses. At the subcellular level, responses in primed cells seem to involve the same stimulus-response coupling pathways that transduce the actions of receptor-mediated stimulation of single agonists. However, the precise mechanism(s) for this priming process remains speculative. In this paper, we briefly review the phenomenon of primed stimulation and focus on the role of PAF as both a priming and stimulating agonist in primed stimulation.

Priming and Primed Stimulation

In the primed cellular response, a priming agonist commonly induces little or no measurable stimulation. It does, however, induce a "primed" state in which cells become hyperresponsive to various other stimuli. For example, bacterial lipopolysaccharide (LPS) by itself fails to elicit responses from PMN. However, LPS primes these cells for marked enhancement of superoxide anion production, lysosomal enzyme release, and adhesiveness upon stimulation with the chemotactic peptide fMLP.[1-3] LPS also primes PMN for enhanced production of the arachidonate metabolite leukotriene B_4 (LTB_4), induced by opsonized zymosan, phorbol myristate acetate (PMA), or the calcium ionophore A23187.[4] In a similar fashion, small concentrations of LPS (< 1mg/ml) have no immediate effects on macrophages but nevertheless prime these cells for augmented arachidonic acid (AA) release and metabolism[5,6] and superoxide production[7] in response to a variety of soluble and particulate stimuli. The protein cytokines tumor necrosis factor-alpha (TNFα)[8] and granulocyte-macrophage colony-stimulating factor (GM-CSF)[9] likewise prime PMN for enhanced superoxide production. The relevance of *in vitro* priming of PMN and macrophages by LPS and inflammatory cytokines is suggested by the observation that PMN obtained from septic patients exhibit enhanced responses when stimulated *in vitro*. These PMN, then, may have been primed *in vivo*.[10,11] Such priming may also play a role in allergic reactions: basophils are primed by interleukin-3 for enhanced release of histamine and leukotrienes.[12]

The intracellular changes induced by priming stimuli that allow for an augmented response have not been clearly defined. Because physiological primers operate through membrane receptors, it is reasonable to hypothesize that priming is mediated through the subcellular pathways involved in stimulus-response coupling. In this schema, agonist receptor binding activates G-proteins within the cell membrane. The G-proteins in turn activate phospholipase(s) C (PLC) and D (PLD). PLC cleaves phosphatidylinositol and other membrane phospholipids to form diacylglycerol (DAG) and inositol phosphates.[13] Inositol triphosphate releases calcium from intracellular stores. The attendant rise in cytosolic calcium promotes the activation of various protein kinases, phospholipase A_2 (PLA_2), and other cellular enzymes. Concurrent production of DAG activates protein kinase C (PKC) by a calcium-facilitated reaction.[14] In addition, activation of PLD releases phosphatidic acid, which is metabolized to DAG.[13] This sequence can induce cellular functions.

The synergistic enhancement of cellular responses that follows priming may occur through partial activation of the above listed pathways just mentioned, resulting in recruitment of additional second messengers by the priming agent. This notion is based on studies that utilize agonists that bypass receptors to selectively increase intracellular calcium (ionophores, for example, A23187) or activate PKC (PMA, DAG). Sequential stimulation with low concentrations of calcium ionophores plus PMA induces marked enhancement of superoxide release in PMN,[15,16] macrophages,[17] and endothelial cells (EC).[18] Synthetic DAG primes PMN responses *in vitro*.[19] LPS may prime cellular responses in a similar manner. Treatment of PMN with LPS increases intracellular calcium concentration, and priming is blocked by intracellular chelation of calcium.[20] Membrane receptors for some soluble PMN agonists are increased in number following LPS exposure.[21] LPS activates PKC and stimulates phosphatidylinositol metabolism in macrophages.[22-24]

Thus, priming by agonist concentrations that fail to stimulate overt cellular responses seems to occur in association with partial activation of stimulus-response coupling pathways.

Stimulus-Response Coupling Mechanisms Involved in PAF-Induced Responses and PAF Production

PAF responses. The bioactions of PAF occur following binding to its cell membrane receptor and appear to be mediated intracellularly by signal transduction pathways. There are two distinct populations of PAF membrane receptors in PMN, a high-affinity receptor (K_d=0.2nM) and a low-affinity receptor (K_d=500nM).[25] These two types of receptors may activate signal transduction pathways differently to induce different cellular responses, depending on the concentration of PAF at the cell membrane.[26] The PAF-receptor complex appears to be linked to guanosine triphosphate (GTP)-binding proteins in PMN. Non-metabolizable GTP analogues, which disrupt G-protein-receptor interactions, down-regulate PAF receptors.[27] Pertussis toxin (PT), which inactivates certain G-proteins by inducing ribosylation, interferes with cellular stimulation by PAF. Pretreatment of PMN with PT inhibits PAF-induced chemotaxis, superoxide production, lysosomal enzyme release, and AA release, suggesting that fully functional G-proteins are necessary for these responses.[28,29]

Stimulation of cells by PAF occurs in association with activation of signal transduction pathways that appear to be dependent on G-proteins. Intracellular calcium concentrations increase in PMN[30] and macrophages[31] following PAF stimulation. The large persistent increase in intracellular calcium in PMN that follows stimulation with concentrations of PAF > 10nM is inhibited by PT, suggesting that low-affinity PAF receptors are linked to G-proteins, whereas the early transient increase in intracellular calcium produced by low concentrations of PAF is unaffected.[26] PAF stimulates phosphoinositide (PI) metabolism in PMN,[26] macrophages,[32] platelets,[33] and EC,[34] which increases intracellular calcium and forms DAG. Such PAF-induced PI turnover in PMN[26] and macrophages[32] is inhibited by PT, which suggests that PI metabolism is mediated by G-protein linked PLC activity.

The increases in intracellular calcium and DAG following PAF-stimulation results in PKC activation in PMN[35] and EC.[34] In addition to facilitating PAF-induced responses, this kinase may also down-regulate some subcellular pathways and cellular functions. Activation of PKC inhibits PAF-induced increases in intracellular calcium and[36] lysosomal enzyme release[26] and reduces the number of high-affinity PAF receptors on the cell membrane.[37] Superoxide production and PI metabolism are inhibited by the PKC activator PMA in PAF-stimulated macrophages as well.[32]

PAF induces the release and metabolism of AA in phorbol myristate acetate (PMN) through a mechanism that depends on G-proteins and increases in intracellular calcium.[38] LTB_4 is produced in

PAF-stimulated PMN.[39] Endogenous AA metabolites may participate in signal transduction by enhancing [hydroxyeicosatetraenoic acids (HETEs)][35] or inhibiting (prostaglandin E) cellular responses.

PAF production. PAF is produced by cells in response to specific stimuli through these same signal transduction mechanisms. The precursor for PAF production is 1-alkyl-2-acyl-*sn*-glycerophosphocholine in cell membranes. Noteably, AA is often present in the 2-acyl position of these lipids so that upon cleavage by PLA_2, AA is released, and the direct precusor for PAF, lyso-PAF, is formed.[40] Activation of PLA_2 is dependent upon increases in intracellular calcium. Calcium ionophores are potent stimuli for PAF production in PMN,[40] macrophages,[41] and EC,[42] as are a variety of other agonists that raise intracellular calcium following receptor binding.

Cellular protein kinases may also participate in PAF production. PKC activation has been reported to facilitate[43] or inhibit[44] PAF production in PMN. Cyclic AMP-dependent protein kinase may enhance PAF production by increasing acetyltransferase activity.[44]

The production of PAF can also be primed. LPS induces small increases in intracellular PAF in PMN but primes for a marked increase in PAF production in response to fMLP or PMA.[45] The lipoxygenase product of AA, 5-HETE, enhances PAF synthesis in response to a nonmetabolizable PAF analogue.[46]

Thus, much evidence indicates that cellular stimulation by PAF is dependent on the activation of several pathways of signal transduction. Also, the synthesis of PAF induced by specific stimuli appears to operate through these same mechanisms of cellular activation. Because other stimuli utilize these same pathways, it seems likely that simultaneous stimulation by PAF and another agonist might modify the cellular response in a synergistic or antagonistic manner.

The Role of PAF in Priming and Primed Stimulation in Isolated Cells *in vitro*

PAF primes. Isolated cells exposed to extremely small concentrations of PAF may become primed for augmented responses upon stimulation with a second agonist. Most studies of PAF priming have been performed using isolated preparations of PMN. PAF by itself induces very little respiratory burst activity in PMN at concentrations less than 100nM.[47] In contrast, PMN incubated with as little as 0.01nM PAF exhibit marked enhancement of superoxide anion production and oxygen consumption when stimulated with fMLP.[47-49] The PAF-primed response can be elicited almost immediately after PAF is added to the cells; this response persists for 60 minutes and is not reversed by cell washing.[47] PAF also primes PMN for enhanced superoxide production in response to PMA but does not affect the response to opsonized zymosan.[47] The PAF-primed response in PMN is not limited to respiratory burst activity. Lysosomal enzyme release and aggregation stimulated by fMLP or PMA are increased in PAF-primed cells.[50] The fMLP-stimulated increase in intracellular calcium and expression of CR3 receptors are enhanced as well.[50] Priming does not occur with lysoPAF or with PAF in the presence of PAF receptor antagonists, indicating that the priming effect requires specific binding of PAF to membrane receptors.[50] It is of interest that if the order of presentation of stimuli is reversed (that is, fMLP before PAF) priming is not observed.[50] This may result from fMLP-induced deactivation of signal transduction pathways. Molski *et al.*[36] have shown that prior exposure to fMLP blocks PAF-induced increases in intracellular calcium through a PT-inhibitable process.

The capacity for PAF to prime cell types other than PMN has not been studied as extensively. Baggiolini *et al.*[51] have reported that monocyte-macrophages are primed by PAF for enhanced superoxide release in response to fMLP. Picomolar concentrations of PAF prime alveolar macrophages for a two- to threefold increase in stimulated TNFα and interleukin 1 (IL1) production.[52] TNFα is also a stimulus for PAF production in these cells, which raises the possibility of a potent positive feedback system within the cell that might augment mediator production.[53]

Endogenous PAF. Studies of PAF priming clearly document that cells exposed to minute concentrations of exogenous PAF may become primed for enhanced responses to other stimuli. Another intriguing question raised by these findings is whether endogenous PAF production primes cells or functions as an intracellular second messenger in signal transduction. Virtually every cell type that synthesizes PAF (PMN, eosinophils, platelets, macrophages, EC) can be stimulated by PAF. Furthermore, with the exception of monocytes, which release 20 to 50 percent of their PAF extracellularly, most cell types (for example, PMN and EC) retain 90 percent or more of newly synthesized PAF within the cell.[54] Based on the parallel stimulation of PAF and superoxide production in LPS-primed PMN, Worthen *et al.*[45] have hypothesized that LPS-primed PAF production in these cells might mediate the generalized priming of cell functions. At this time, the role for PAF as an endogenous primer is not firmly established, but it is interesting to speculate such a function for the large amounts of PAF that are produced and retained within stimulated inflammatory effector cells.

Priming of PAF responses. Cellular responses induced by exogenous PAF may be primed for enhanced activity by other agonists. In contrast to the findings of Vercellotti *et al.*,[50] Dewald and Baggiolini[48] demonstrated that PAF did not induce superoxide anion release from control PMN, but did cause a modest response in fMLP-primed cells. Similarly, PMN pretreated with cytochalasin B exhibit markedly enhanced release of superoxide anion and lysosomal enzymes and respond to PAF concentrations that are several logfold less.[49]

High concentrations of the PKC activator PMA down-regulate high-affinity PAF receptors and block the effects of exogenous PAF on PMN. On the other hand, lower concentrations of PMA actually prime PMN for enhanced degranulation and PKC activation in response to PAF.[37] DAG, which also directly activates PKC, increases the potency of PAF 10- to 30-fold and enhances the magnitude of PAF-induced lysosomal enzyme release in PMN.[55] These findings suggest that submaximal PKC activation may synergize with other receptor-mediated signal transduction pathways to augment functional responses to PAF.

The major pathway of AA metabolism in PMN is through lipoxygenases, producing HETEs and leukotrienes. O'Flaherty and Nishihira[35] have shown that exposure of PMN to exogenous 5-HETE enhances PKC activation in response to PAF and increases the potency of PAF in degranulation by 100- to 1000-fold.[56] In addition, 5-HETE primes PMN for enhanced PAF production in response to nonmetabolizable PAF analogues.[46] These findings suggest that endogenous 5-HETE, produced in response to other agonists may augment PAF-induced responses in PMNs. The protein cytokine GM-CSF primes PAF-stimulated increases in intracellular calcium in PMN which might enhance cellular responses to PAF.[57]

In summary, these findings indicate that PAF synergizes with a variety of other protein and lipid mediators to augment responses in purified inflammatory cells *in vitro*. Exogenous PAF effectively primes these cells and induces enhanced responses in cells primed with other agonists. Endogenous PAF produced following stimulation may function intracellularly to amplify cellular responses. Taken together, this evidence suggests that PAF could have similar effects during inflammation *in vivo,* where effector cells are bombarded with an array of mediators.

Priming and Primed Stimulation in Whole Animals

Synergistic interactions among mediators may explain the explosive injury that occurs *in vivo* in the setting of septic shock, wherein multiple inflammatory mediators are produced locally and systemically.[58] PAF in particular may be a pivotal mediator in sepsis. PAF is released into the circulation of animals given LPS.[59] Furthermore, PAF receptor antagonists reduce the adverse cardiovascular and pulmonary effects of endotoxemia and prolong survival in animals given lethal doses of LPS.[59-61] Based on its ability to mediate priming and primed stimulation *in vitro,* PAF may augment the effects of other stimuli *in vivo.*

Defining the role of specific mediators of priming is much more difficult *in vivo* because of the multiplicity of cell types and agonists involved in the inflammatory response in the whole organism. Studies in experimental animals have addressed this problem. The classic description of priming *in vivo* is the Shwartzman reaction, in which sequential administration of innocuous doses of LPS induce an explosive inflammatory reaction.[62] A similar synergistic inflammatory response results from coadministration of TNFα and IL 1 in concentrations that produce minimal inflammation when administered alone.[63] Studies by Worthen *et al.*[64] and Haslett *et al.*[65] have shown that small doses of LPS and the chemotactic peptide fNLP synergistically induce prolonged PMN sequestration in the lung and edematous lung injury in whole rabbits. Similarly, rabbits given benign doses of intravenous PAF experience more profound and prolonged neutropenia following fMLP administration compared with animals given FMLP alone.[66] PAF also synergizes with LPS to produce ischemic bowel necrosis in rats. In this model, administration of LPS and PAF together in doses that fail to induce injury by themselves results in necrotic lesions in the intestines.[67] Further studies revealed that bowel necrosis could be induced by synergistic stimulation with TNFα and LPS and was inhibited by PAF receptor antagonists. Individually, both TNFα and LPS induced PAF production within the bowel wall, and when given together, this effect was additive.[68] In summary, these studies demonstrate that synergistic stimulation is elicited by a variety of agonists in different animal models to produce inflammatory injury, suggesting that the primed response occurs *in vivo*.

The Rabbit Isolated Perfused Lung Model for Primed Lung Injury

In the whole animal, synergistic stimulation occurs in a manner similar to *in vitro* priming of purified populations of cells. However, the pathophysiologic mechanisms involved in this response are much more difficult to dissect because of complex interactions between multiple secondary mediators and cell types. Our interest has been the role of multiple mediators in induction of acute inflammatory lung injury. In order to study the possibility of synergistic stimulation and priming by multiple mediators in lung injury, we have used the isolated buffer-perfused rabbit lung (IPL). In this system, the pulmonary circulation is cannulated, and the heart and lungs are removed from the animal.[69] The pulmonary circulation is washed free of serum and circulating blood cells and then perfused with physiologic buffer. This allows us to study the direct effects of agonists on the lung itself in the absence of mediators derived from serum (comple-

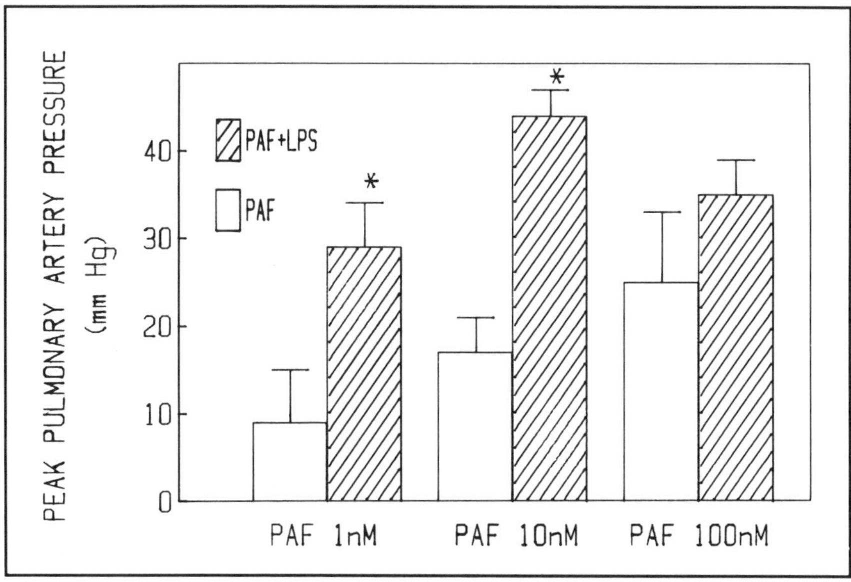

Fiaure 1. Mean peak pulmonary artery pressure in rabbit IPLs that were perfused for two hours with LPS-free buffer (<100pg/ml LPS) (PAF) or buffer containing 100ng/ml of LPS (PAF+LPS) and then stimulated with PAF. Data are mean ± S.E.M.
*indicates those values that are significantly different ($P < 0.05$) for PAF versus LPS + PAF.

ment components) and circulating blood cells (PMN, platelets, monocytes).

We have used the rabbit IPL to determine whether combinations of inflammatory mediators synergize to amplify lung injury.[69] In the IPL, perfusion with buffer containing 100ng/ml of LPS for two hours induces slight elevations in pulmonary artery pressure (PAP) compared with lungs perfused with LPS-free buffer. LPS at this dosage does not cause lung edema as detected by changes in lung weight. Administration of 1nM, 10nM, or 100nM PAF induced mild pulmonary hypertension with peak PAPs of 9 ± 5, 17 ± 4, and 25 ± 8 mm Hg, respectively, but little edema in lungs perfused with LPS-free buffer for two hours [baseline PAP ≈ 5 to 10 mm Hg]. In contrast, the PAF-induced response is markedly enhanced in lungs perfused for two hours with LPS prior to PAF administration. The mean peak PAPs in response to 1nM PAF (29 ± 5mm Hg) and 10nM PAF (44 ± 3mm Hg)

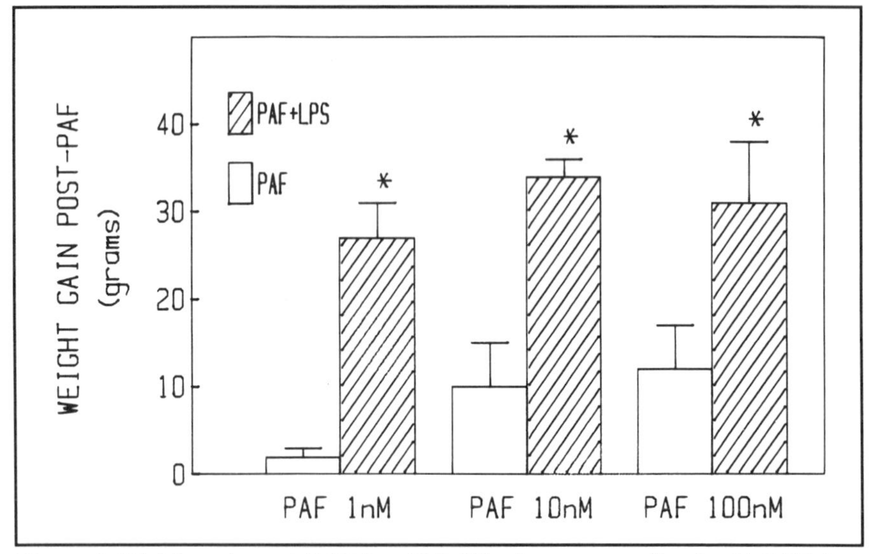

Figure 2. Weight gain post-PAF in rabbit IPLs that were perfused with LPS-free (PAF) or LPS (100ng/ml) (PAF + LPS) buffer for two hours and then stimulated with PAF. Data are mean ± S.E.M. of weight gain in the 60 minutes following PAF administration.
*indicates those values that are significantly different ($P < 0.05$) for PAF versus LPS + PAF.

were significantly increased in LPS-primed lungs (Figure 1). Significant pulmonary edema (measured as weight gain) developed in LPS-primed lungs after PAF stimulation (Figure 2). These findings indicate that LPS primes the rabbit IPL, making it susceptible to severe injury in response to amounts of PAF that are relatively benign in unprimed lungs.

In whole animals, cyclooxygenase products of AA appear to mediate pulmonary hypertension and edema in the early phases of LPS-induced lung injury.[70] In addition, PAF stimulates the production of AA metabolites in the guinea pig IPL.[71] We therefore studied the role of AA metabolites in LPS-primed, PAF-induced injury in the IPL. The cyclooxogenase inhibitor, indomethacin, completely blocked the LPS-primed response.[69] Pretreatment within 10μM indomethacin prevented lung edema in LPS-primed, PAF-stimulated IPLs and completely inhibited the enhanced PAP response. Mean peak PAPs in LPS-primed lungs given indomethacin were similar to those induced

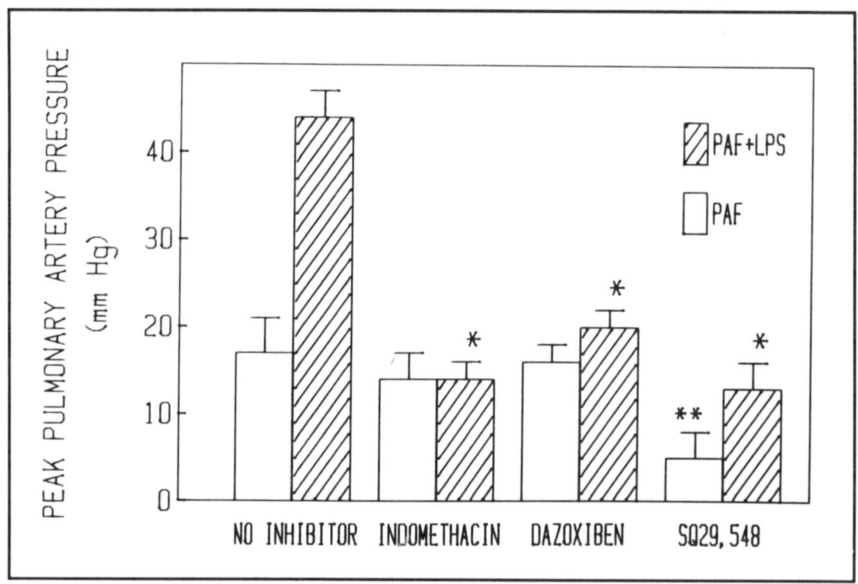

Figure 3. Effect of inhibitors on PAF-induced pulmonary hypertension. Rabbit IPLs were perfused for two hours with LPS-free (PAF) or LPS (100ng/ml) (LPS+PAF) buffer and then stimulated with 10nM PAF. IPLs were treated with no inhibitor, 10μM indomethacin, 5μM dazoxiben, or 1μM SQ 29,548. Data are mean ± S.E.M.
*indicates those values that are significantly different ($P < 0.05$) for IPLs treated with PAF+LPS and an inhibitor vs. PAF+LPS with no inhibitor.
**indicates a significant difference ($P < 0.05$) between PAF and LPS+PAF treated with SQ 29,548 treatment.

by PAF in unprimed lungs (Figure 3). We then measured the production of thromboxane by determining the levels of its stable metabolite thromboxane B_2 (TxB_2) in the perfusate of the IPL. Indomethacin (10μM) completely blocked the production of TxB_2 induced by PAF in both LPS primed and unprimed lungs. In unprimed lungs, PAF induced a rapid rise in perfusate TxB_2 to 180 percent of pre-PAF levels at three minutes. TxB_2 rapidly returned to near baseline levels at five minutes. LPS priming enhanced both the magnitude and duration of TxB_2 production. Three minutes after PAF administration, TxB_2 was 264 percent of pre-PAF levels and remained elevated for the duration of the experiment. These findings indicate that LPS priming markedly enhanced PAF-stimulated TxB_2 production in the IPL. The time courses of PAF-induced TxB_2 production and elevation of PAP were very similar, suggesting that enhanced thromboxane production might be responsible for the LPS-primed response.[69]

To determine the role of thromboxane production in LPS-primed injury in this model, we used the thromboxane synthase inhibitor dazoxiben and the thromboxane receptor antagonist SQ 29548. Both dazoxiben (5μM) and SQ 29,548 (1μM) completely blocked the LPS-primed response in the PAF-stimulated IPL (see Figure 3).[69] These findings indicate that augmentation of PAF-induced thromboxane production is responsible for enhanced injury in the LPS-primed IPL. The findings also demonstrate that the lung itself can be primed by LPS for synergistic enhancement of inflammatory injury in the absence of circulating blood cells and plasma mediators. The similarities between LPS-induced injury in the whole animal[70] and LPS-primed injury in the IPL suggest that the IPL may provide a valuable model of septic lung injury.

Conclusions

Multiple mediators synergize in isolated organs and cell preparations *in vitro* and in whole animals to amplify the production of autotoxins to enhance inflammatory injury. PAF may participate as a primary, intermediary, and perhaps terminal agonist in these responses. It may indeed occupy an essential role in overwhelming inflammatory states *in vivo* (anaphylaxis, sepsis syndrome, and adult respiratory distress syndrome). Determining the precise mechanisms and cells responsible for priming may provide important information regarding the pathogenesis of inflammation and will likely improve our ability to design novel therapeutic modalities. Successful intervention may require interruption of several primary processes, because multiple mediators may be responsible for the final injury.

References

1. L. Guthrie *et al.*, *J. Exp. Med.* **160**, 1656 (1984).
2. C. Haslett *et al.*, *Am. J. Pathol.* **119**, 101 (1985).

3. C. Dahinden and J. Fehr, *J. Immunol.* **130**, 863 (1983).
4. M. Doerlier *et al., J. Clin. Invest.* **83**, 970 (1989).
5. A. Aderem *et al., J. Exp. Med.* **164**, 165 (1986).
6. A. Aderem and Z. Cohn, *J. Exp. Med.* **167**, 623 (1988).
7. M. Pabst and R. Johnston, *J. Exp. Med.* **151**, 101 (1980).
8. R. Berkow *et al., J. Immunol.* **139**, 3783 (1987).
9. R. Weisbart *et al., Blood* **69**, 18 (1987).
10. C. McCall *et al., J. Infect. Dis.* **140**, 277 (1979).
11. G. Zimmerman, A. Renzetti, and H. Hill, *Am. Rev. Respir. Dis.* **127**, 290 (1983).
12. Y. Kurimoto, A. DeWeck, and C. Dahinden, *J. Exp. Med.* **170**, 467 (1989).
13. J. Exton, *FASEB J.* **2**, 2670 (1988).
14. U. Kikkawa and Y. Nishizuka, *Ann. Rev. Cell Biol.* **2**, 149 (1986).
15. L. McPhail, C. Clayton, and R. Snyderman, *J. Biol. Chem.* **259**, 5768 (1984).
16. J. Robinson *et al., Biochem. Biophys. Res. Commun.* **122**, 734 (1984).
17. T. Finkel *et al., J. Biol. Chem.* **262**, 12589 (1987).
18. T. Matsubara and M. Ziff, *J. Cell. Physiol.* **127**, 207 (1986).
19. D. Bass *et al., J. Biol. Chem.* **264**, 19610 (1989).
20. J. Forehand *et al., J. Clin. Invest.* **83**, 74 (1989).
21. D. Goldman *et al., J. Immunol.* **137**, 1971 (1986).
22. P. Wightman and C. Raetz, *J. Biol. Chem.* **259**, 10048 (1984).
23. R. Zoeller *et al., J. Biol. Chem.* **262**, 17212 (1987).
24. V. Pripic *et al., J. Immunol.* **139**, 526 (1987).
25. J. O'Flaherty *et al., J. Clin. Invest.* **78**, 381 (1986).
26. P. Naccache *et al., J. Leukocyte Biol.* **40**, 533 (1986).
27. D. Ng and K. Wong, *Biochem. Biophys. Res. Commun.* **141**, 353 (1986).
28. P. Lad, C. Olson, and I. Grewal, *Biochem. Biophys. Res. Commun.* **129**, 632 (1985).
29. P. Lad *et al., Proc. Natl. Acad. Sci. USA* **82**, 8643 (1985).
30. V. von Tscharner *et al., Nature (London)* **324**, 369 (1986).
31. G. Conrad and T. Rink, *J. Cell Biol.* **103**, 439 (1986).
32. S. Huang *et al., Biochem. J.* **249**, 839 (1988).
33. M. Billah and E. Lapetina, *Proc. Natl. Acad. Sci. USA* **80**, 965 (1983).
34. G. Grigorian and U. Ryan, *Circ. Res.* **61**, 389 (1987).
35. J. O'Flaherty and J. Nishihira, *J. Immunol.* **138**, 1889 (1987).
36. T. Molski *et al., Biochem. Biophys. Res. Commun.* **151**, 836 (1988).
37. J. O'Flaherty, D. Jacobson, and J. Redman, *J. Biol. Chem.* **264**, 6836 (1989).
38. S. Nakashima *et al., J. Immunol.* **143**, 1295 (1989).
39. A. Lin, D. Morton, and R. Gorman, *J. Clin. Invest.* **70**, 1058 (1982).
40. F. Chilton *et al., J. Biol. Chem.* **259**, 12014 (1984).
41. M. Elstad *et al., J. Immunol.* **140**, 1618 (1988).
42. G. Camussi *et al., J. Immunol.* **131**, 2397 (1983).
43. T. McIntyre *et al., J. Biol. Chem.* **262**, 15370 (1987).
44. M. Nieto, S. Velasco, and M. Crespo, *J. Biol. Chem.* **263**, 4607 (1988).
45. G. Worthen *et al., J. Immunol.* **140**, 3553 (1988).
46. T. Tessner, J. O'Flaherty, and R. Wykle, *J. Biol. Chem.* **264**, 4749 (1989).
47. J. Gay *et al., Blood* **67**, 931 (1986).
48. B. Dewald and M. Baggiolini, *Biochem. Biophys. Res. Commun.* **128**, 297 (1985).
49. L. Ingraham *et al., Blood* **59**, 1259 (1982).
50. G. Vercellotti *et al., Blood* **71**, 1100 (1988).

51. M. Baggiolini, B. Dewald, and M. Thelan, *Prog. Biochem. Pharmacol.* **22,** 90 (1988).
52. C. Dubois, E. Bissonnette, and M. Rola-Pleszczynski, *J. Immunol.* **143,** 964 (1989).
53. G. Camussi *et al., J. Exp. Med.* **166,** 1390 (1987).
54. J. Lynch and P. Henson, *J. Immunol.* **137,** 2653 (1986).
55. J. O'Flaherty *et al., Biochem. Biophys. Res. Commun.* **123,** 64 (1984).
56. Ibid. **111,** 1 (1983).
57. P. Naccache *et al., J. Immunol.* **140,** 3541 (1988).
58. N. Olson, W. Salzer, and C. McCall, *Molec. Aspects Med.* **10,** 511 (1988).
59. S. Chang *et al., J. Clin. Invest.* **79,** 1498 (1987).
60. T. Doebber *et al., Biochem. Biophys. Res. Commun.* 127, 799 (1985).
61. Z. Terashita *et al., Eur. J. Pharmacol.* **109,** 257 (1985).
62. E. Gaynor, C. Bouvier, and T. Spaet, *Science* **170,** 986 (1970).
63. H. Movat *et al., Am. J. Pathol.* **129,** 463 (1987).
64. G. Worthen *et al., Am. Rev. Respir. Dis.* **136,** 19 (1987).
65. C. Haslett *et al., Am. Rev. Respir. Dis.* **136,** 9 (1987).
66. L. Ingraham *et al., Br. J. Haematol.* **66,** 219 (1987).
67. F. Gonzalez-Crussi and W. Hsueh, *Am. J. Pathol.* **112,** 127 (1983).
68. X. Sun and W. Hsueh, *J. Clin. Invest.* **81,** 1328 (1988).
69. W. Salzer and C. McCall, *J. Clin. Invest.* **85,** 1135 (1990).
70. K. Brigham and B. Meyrick. *Am. Rev. Respir. Dis.* **133,** 913 (1986).
71. Y. Hamasaki *et al., Am. Rev. Respir. Dis.* **129,** 742 (1984).

Clinical Applications for PAF

11

Involvement of Platelet-Activating Factor in Gastrointestinal Disease

Brendan J.R. Whittle
Department of Pharmacology
Wellcome Research Laboratories
Langley Court, Beckenham
Kent BR3 3BS
United Kingdom

PAF is a potent ulcerogen of the gastrointestinal tract, inducing acute mucosal injury in the stomach and small intestine. The profile of the pathophysiological events following administration of PAF, including intense vascular and cellular changes, resemble those of the clinical state of shock. Such a range of biological properties therefore make PAF a likely candidate as a pathological mediator of gastrointestinal disorders in conditions of shock and inflammation. Indeed, increased gastrointestinal levels of PAF can be detected during endotoxemia, while PAF-receptor antagonists can ameliorate the gastrointestinal damage in such conditons of shock. Likewise, gastrointestinal motility induced by PAF and endotoxin is inhibited by PAF-receptor antagonists, and such agents may be of benefit in therapy of motility dysfunction. Endogenous PAF may also be involved in chronic intestinal disorders such as inflammatory bowel disease. Thus, pharmacological agents that modify the actions of PAF may be of clinical value for a number of gastrointestinal utilities.

Introduction

PAF-acether (platelet-activating factor, PAF) induces a range of pathophysiological events that resemble shock states.[1,2] These actions include increased vascular permeability, hemoconcentration, hypotension and circulatory collapse, neutrophil aggregation, and lysosomal enzyme release.[3] This profile of the biological actions of PAF thus makes it a potential candidate as a mediator of gastrointestinal damage in conditions of both shock and inflammation. This chapter reviews the nature of the damage induced by PAF, its role in the pathogenesis of gastric and intestinal disease, and the therapeutic potential of PAF antagonists in these conditions.

Effects on the Gastric Mucosa

Effects of intravenously administered PAF in the rat. Intravenous infusion of PAF in low doses induced extensive damage to the rat gastric mucosa, results that have shown PAF to the most potent gastric ulcerogen so far described.[4] The mucosal damage was macroscopically characterized as extensive hyperemia and vasocongestion, with areas of hemorrhagic damage. On histological evaluation, microvascular engorgement and congestion were with epithelial and glandular destructions the predominant features, with focal areas of necrosis that extended throughout the depth of the mucosal and submucosal microvasculature. The presence of white cell aggregates within these vessels was also observed, particularly at the base of the mucosal glands and submucosal vessels.[4]

Close-arterial administration of PAF in the rat. Local intraarterial infusion of PAF resulted in a dose-dependent increase in macroscopically assessed damage to the rat gastric mucosa (Figure 1). The macroscopic appearance of this damage was comparable to that described following intravenous infusion and was characterized by extensive hyperemia and hemorrhage located in the corpus region.[5]

The threshold local doses of PAF for induction of macroscopic or histological damage in both antral and corpus regions following

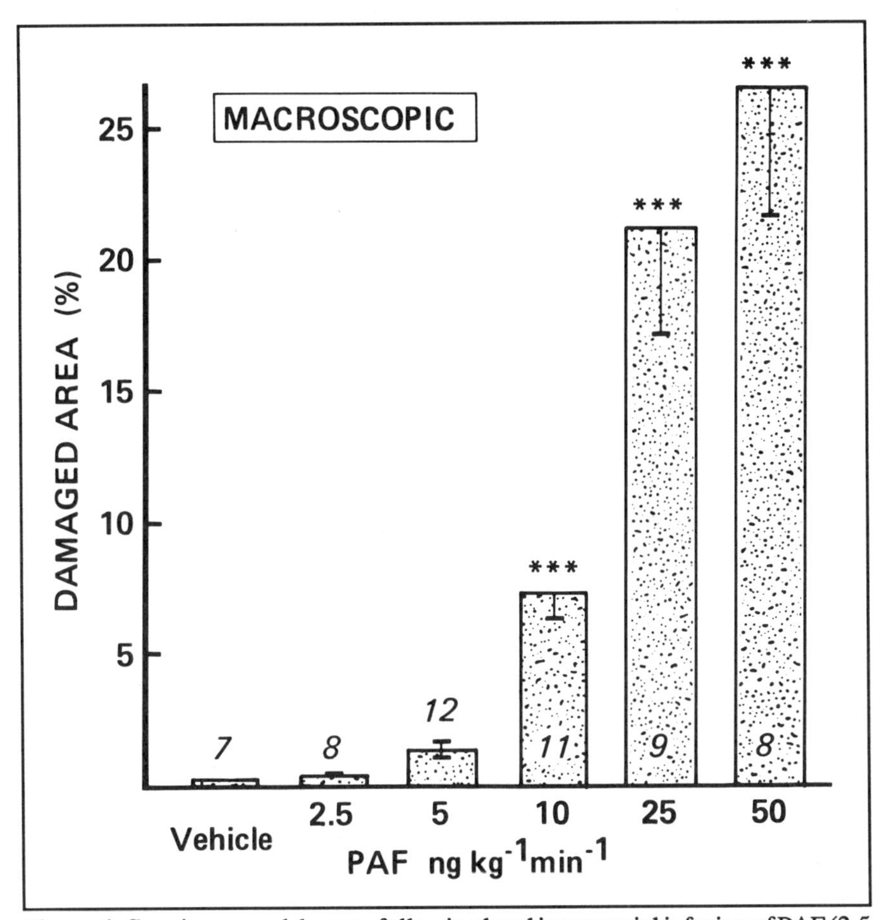

Figure 1. Gastric mucosal damage following local intraarterial infusion of PAF (2.5 to 50ng kg^{-1} min^{-1}) for 10 minutes in rat. Area of macroscopically assessed damage was determined by computerized planimetry and expressed as percent of total mucosal area. Results are shown as mean ± S.E.M. of (*n*) studies, where statistical difference from vehicle control group is given as $^{***}P < 0.001$. Data adapted from Reference 5.

close-arterial infusion produced minimal systemic hypotension, which indicated a dissociation between systemic cardiovascular actions and gastric damage.[5] The predominant histological feature of congestion and engorgement of the mucosal microvessels following both local and intravenous administration of PAF suggests that vascular stasis may be an important mechanism underlying the mucosal damage (Figure 2).

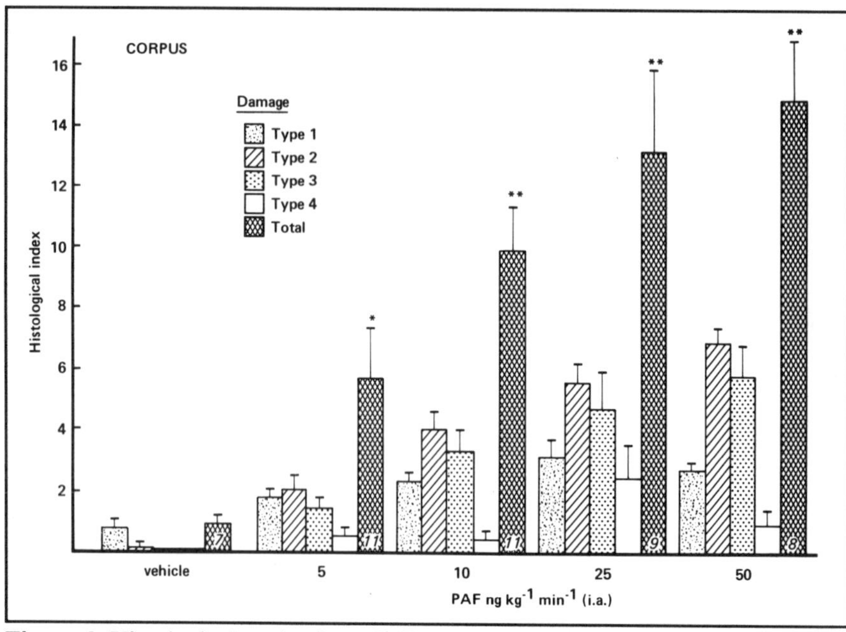

Figure 2. Histological evaluation of effects of local intraarterial infusion of PAF-acether (5 to 50ng kg^{-1} min^{-1}) on gastric corpus mucosa of rat. Data are shown in terms of histological score for various types of damage (types 1 to 4). Thus, epithelial cell damage was type 1 (a score of 1); glandular disruption, vasocongestion, or edema in upper mucosa (type 2); hemorrhagic damage in the mid to lower mucosa (type 3); and deep necrosis or ulceration (type 4). Each subsection was evaluated on a cumulative basis, and overall mean value of scores for each of five to six fields was taken as histological index for that section. Results are shown as mean ± S.E.M. of (n) values. For clarity, significant differences from vehicle control are given for total histological index only, where $*P < 0.05$, $**P < 0.01$. Data adapted from Reference 5.

Microvascular actions of PAF in the rat. To investigate whether the microcirculatory stasis could have resulted from direct vasoconstriction in the submucosal arterioles and venules, an *in vivo* microscopy technique in the rat has been utilized.[6] No change in the diameter of these submucosal vessels could be detected during intravenous administration of PAF in doses that induced mucosal damage.[6] However, slowing of blood flow could clearly be observed during administration of PAF, with stasis of flow being observed in over 50 percent of these microvessels.

To investigate whether changes in blood flow could be detected

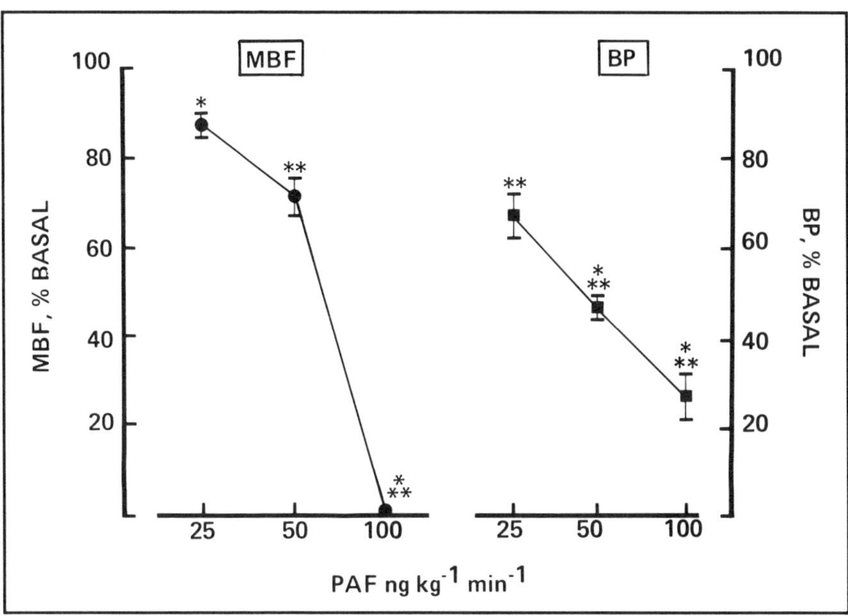

Figure 3. Effect of intravenous infusion of PAF (25 to 100ng $kg^{-1}min^{-1}$ on gastric mucosal blood flow (MBF), determined by hydrogen-gas clearance and mean systemic arterial blood pressure (BP) in rat. Results, expressed as percent of basal values, are mean ± S.E.M. of four to seven experiments in each group. MBF was measured four minutes after start of PAF infusion. Statistical difference from control data is shown as $*P < 0.05$, $**P < 0.01$, $***P < 0.001$. Data adapted from Reference 6.

in the mucosal capillaries, red blood cell velocity was determined, again using *in vivo* microscopy techniques.[6] Intravenous infusion of PAF-acether induced a significant fall in capillary blood flow within two to three minutes of the start of infusion, with sluggish flow and sludging of blood cells along the capillary walls being observed. At the highest dose of PAF used, capillary blood flow fell to nondetectable levels within five minutes of the start of infusion, and no flow could be detected during the remainder of the infusion period.[6] Likewise, gastric mucosal blood flow determined by hydrogen-gas clearance was substantially reduced following PAF infusion[6] as shown in Figure 3.

Effects in the canine gastric mucosa. The effects of intravenous or local gastric intraarterial infusion of PAF has also been investigated

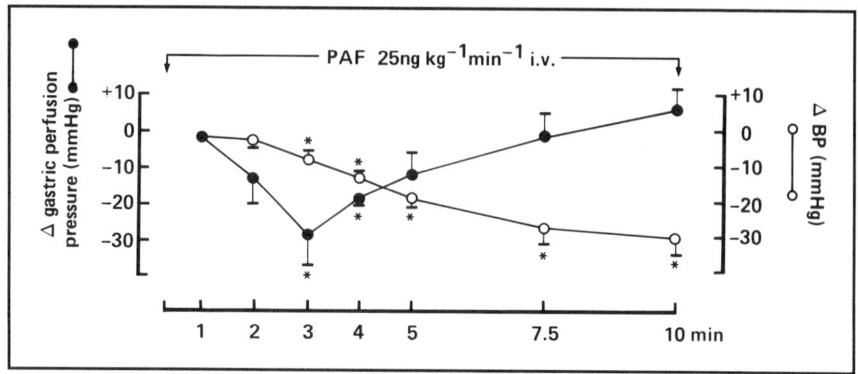

Figure 4. Effect of intravenous infusion of PAF (25ng kg^{-1} min^{-1} for 10 minutes) on (1) gastric vascular perfusion pressure (●) as a measure of vascular resistance and (2) systemic arterial blood pressure (○) in the anesthetized dog. Initial fall in vascular resistance was followed by rebound to values greater than the control, while BP gradually fell during the period of PAF infusion. Macroscopically, initial flushed appearance of the mucosa was followed by a darkened vasocongested appearance. Results, shown as mmHg, are mean ± S.E.M. of three experiments, where significant difference from control is shown as $*P < 0.05$. Data adapted from Reference 7.

in the dog, utilizing the chambered gastric mucosa *in situ.*[7] During intravenous infusion of PAF, an initial fall in gastric vascular resistance was followed by a rise above the resting value (Figure 4), an effect that was also observed following local intraarterial infusion. Following termination of the PAF infusion, vascular resistance continued to increase, accompanied by a darkened vasocongested appearance of the mucosal surface. On histological examination, extensive vascular congestion extended into the submucosa, with extravasation of erythrocytes and prominent subepithelial edema and hemorrhage. Other researchers have also confirmed that intraarterial infusion of PAF reduced gastric venous outflow and mucosal blood flow in the dog.[8]

Mechanisms of mucosal damage. The gastric ulcerogenic effects of PAF are not simply nonspecific actions of a phospholipid because lysoPAF in doses some 100-fold greater than those of PAF failed to induce detectable mucosal damage. Pharmacological evidence that these effects are mediated through activation of specific PAF receptors was provided by the observations that intravenous administration of the structurally dissimilar PAF-receptor antagonists CV-3988, BN-52021, RO-193704, or L-652,731 could inhibit the damage induced by PAF-infusion.[5,9] Other researchers have also reported that the gastric mucosal damage induced by bolus injections of PAF could be attenuated by the PAF-receptor antagonists WEB 2086, BN 52063, and BN 52021.[10,11]

Deposition of blood cells along the walls of the rat gastric microvessels was observed during the periods of sluggish blood flow in the gastric microcirculation induced by PAF.[6] These cells were not identified as platelets, and indeed PAF is known to be only weakly active in stimulating platelet aggregation in the rat. Furthermore, extensive mucosal damage can be induced by PAF in platelet-depleted rats.[4]

The presence of neutrophil aggregates was, however, observed on histological examination of the rat gastric mucosa, and systemic neutropenia has been detected after infusion of proulcerogenic doses of PAF.[12] These aggregates, by initially occluding the smaller capillaries, may thus contribute to the reduction in mucosal blood flow. Such effects would be enhanced by the concurrent, marked hemoconcentration resulting from the extensive systemic extravasation of plasma protein that follows administration of PAF given in ulcerogenic doses. Thus, the changes in the rheological properties of the blood as a consequence of such hemoconcentration, coupled with the presence of aggregated cells, could account for the sluggish flow and eventual stasis observed in the gastric microcirculation. In the dog, the pronounced changes in the gastric microcirculatory parameters are also suggestive of direct vascular spasm or the accumulation of microaggregates in the mucosal microvasculature, thus reducing blood flow. Damage to the feline gastric mucosa characterized as vasocongestion and subepithelial edema is also observed following intravenous infusion of PAF.[13]

PAF in models of gastric ulceration. The involvement of PAF in the mechanisms underlying gastric mucosal damage induced by agents other than endotoxin is not clear. Thus, whereas neither R0-193704 or BN 52021 reduced the hemorrhagic necrotic damage induced by oral administration of ethanol, CV-3988 did exert a protective action, although this action is thus likely to reflect a nonspecific activity.[9] Other researchers have observed that BN 52021 and kadsurenone exerted inconsistent actions on mucosal damage induced by ethanol, aspirin, or restraint stress.[11,14] It is therefore unlikely that PAF release makes a significant contribution to the pathogenesis of these latter models of gastric ulceration.

Intestinal Actions of PAF

The intestinal ulcerogenic actions of an intravenous infusion of PAF-acether have been studied in the rat.[15] Damage to the duodenum, jejunum, and ileum was assessed both histologically and using intraluminal acid phosphatase release as a marker of cellular damage. PAF infusion caused extensive hemorrhagic damage to each of the regions examined, while acid phosphatase release was significantly elevated in the stomach, jejunum, ileum, and duodenum.[15] Likewise, PAF induced plasma leakage in these regions of the gastrointestinal tract. A similar profile of damage could be elicited by intravenous administration of endotoxin from *Escherichia coli*.[9] Other researchers have demonstrated that a combination of endotoxin and high bolus doses of PAF induce a necrotizing enterocolitis in the rat jejunum.[16]

Release of PAF From Gastrointestinal Tissue

The formation of PAF by segments of the rat gastrointestinal tract has been demonstrated following extraction and thin-layer chromatography, by specific bioassay utilizing its potent ability to aggregate suspensions of rabbit washed platelets.[17]

Endotoxin has been reported to increase the formation of PAF in different tissues and blood.[18,19] Likewise, intravenous injection of *E. coli* endotoxin led to a time-dependent increase in the jejunal formation of PAF, which after 20 minutes was 20-fold greater than the control level (Figure 5). There was a significant correlation between

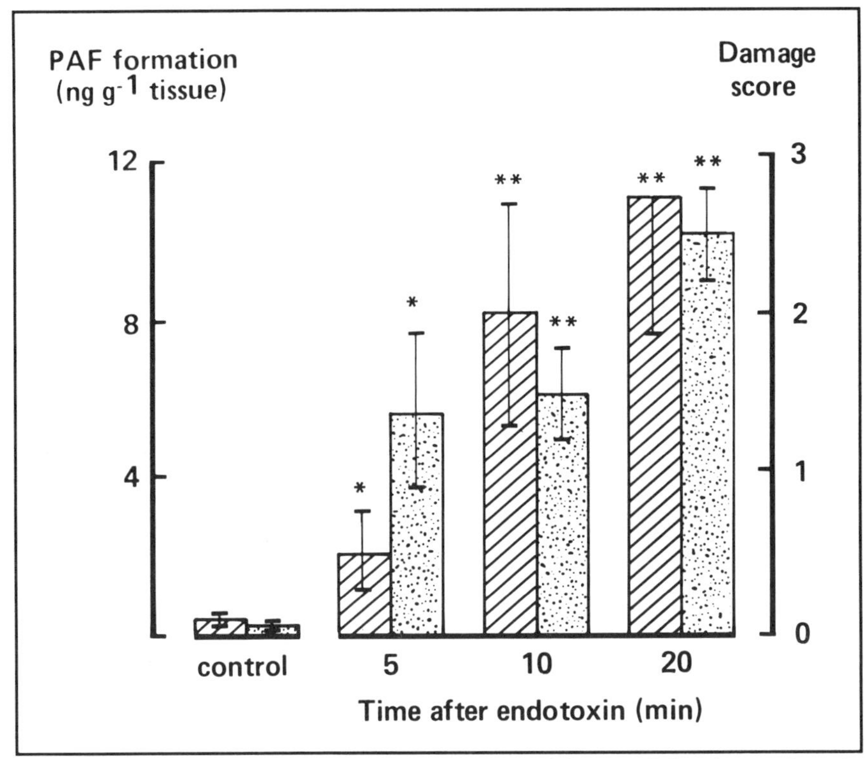

Figure 5. Increase in formation of PAF by rat jejunum following endotoxin administration and its relationship to intestinal damage. Jejunal damage was scored from 0 to 3 depending on severity (speckled columns), and levels of PAF were measured by bioassay (hatched columns) and assessed at various times after administration of *E. coli* lipopolysaccharide (50mg kg^{-1} I.V.). Results are expressed as mean ± S.E.M. of 9 to 12 rats in each experimental group where statistical difference from control is shown as $*P < 0.05$, $**P < 0.01$. Data derived from Reference 17.

elevated PAF release and intestinal hyperemia and hemorrhage, thus supporting a role for PAF as a mediator of such damage.[17]

The increased levels of PAF did not appear to reflect a nonspecific stimulation of phospholipase A_2 and liberation of precursor lipids following damage to the tissue, because the jejunal formation of prostaglandin E_2 was not enhanced.[20] However, jejunal thromboxane B_2 levels, as determined by radioimmunoassay, were significantly increased following endotoxin challenge, and the intestinal damage could be attenuated by thromboxane synthase inhibitors.[20] In that study, pretreatment with dexamethasone also reduced both endotoxin-induced PAF and thromboxane formation by jejunal tissue, while inhibiting the severity of the tissue damage. These findings indicate a complex interaction or sequential release of these tissue destructive mediators underlying the intestinal damage in this model of endotoxemia.

Because the local release of PAF may be involved in the acute or chronic phases of inflammatory conditions in the gastrointestinal tract, the colonic formation of PAF in models of colitis in the guinea pig and rat has been investigated In guinea pigs previously sensitized to ovalbumin, subsequent anaphylactic challenge significantly increased colonic PAF levels.[21] In an immune model of colitis in guinea pigs induced by skin sensitization and subsequent intracolonic challenge with dinitrochlorobenzene, the PAF levels in inflammed tissue were elevated twofold, 24 hours after challenge, when gross hyperemia and edema accompanied by inflammatory cell infiltration were observed. Leukocyte infiltration, as quantitated by myeloperoxidase activity (MPO), was also increased.[21]

In a chronic model of inflammatory bowel disease in the rat induced by intracolonic application of trinitrobenzene sulfonic acid (TNB), colonic PAF levels were elevated by eightfold 24 hours after challenge, with an elevenfold increase in MPO activity.[21] The levels of PAF remained elevated one week after TNB challenge, as was the MPO activity. In a more chronic study using this TNB model of colitis, elevated colonic levels of PAF could be detected one to three weeks after initial challenge, whereas the PAF antagonist BN 52021 administered during the first or second week, reduced the degree of colonic damage.[22]

In studies of human colonic tissue, increased levels of PAF have also been detected in biopsies from patients with active colitis, following *in vitro* challenge with the ionophore A23187 and anti immunoglobulin E; these increased PAF levels could be inhibited by *in vitro* incubation with the clinically effective agents, sulfasalazine, 5-amino-salicylic acid, or prednisolone.[23]

Actions of PAF-Receptor Antagonists

PAF receptor antagonists reduce the systemic hypotension induced by endotoxin in the rat.[3,19,24,25] Likewise, studies with several structurally unrelated PAF antagonists, CV-3988, BN 52021, and RO-193704, have demonstrated substantial reduction in endotoxin-induced damage both to the gastric and intestinal mucosa.[9,26] The intestinal damage induced by bolus injections of PAF in combination with lipopolysaccharide damage in the rat was also prevented by pretreatment with the PAF-receptor antagonists, SR1 63-072, SR1 63-119, or ONO-6240.[27,36] Other workers have also demonstrated that BN 52021 could reduce the gastric and small intestinal disruption induced by *Salmonella enteritidis*.[11] These studies, therefore, support the hypothesis that PAF is an important mediator of the systemic hypotension and intestinal plasma leakage observed during endotoxic shock and that its endogenous release may contribute to the gastrointestinal damage and ulceration associated with this syndrome. Thus, PAF-receptor antagonists may be clinically useful for prevention of such ulceration.

PAF can exert potent spasmogenic actions on segments of gastrointestinal smooth muscle *in vitro*. Thus, PAF induced long-lasting contractions of isolated strips of guinea pig ileum.[28-30] These contractile actions on guinea pig ileum appear to be receptor-mediated because they were prevented by the PAF antagonists. Furthermore, specific binding sites for PAF could be detected on purified plasma membranes from ileal smooth muscle.[31]

PAF also induced dose-related contractions of isolated segments of rat duodenum, jejunum, ileum, forestomach, and colon, with higher concentrations inducing a biphasic contractile response in the small intestinal tissues.[32,33] More recent studies on the rat isolated colon indicated that the potent spasmogenic actions of PAF were inhibited by the PAF-receptor antagonists FR-900452 and CV-3988 but not by antagonists of histamine, 5-hydroxytryptamine, or leukotriene D_4.[34]

PAF, like endotoxin, also stimulates gastric motility *in vivo* by direct actions on the smooth muscle, by histamine release, and by local noncholinergic nonadrenergic neuronal activity.[35] These effects were abolished by PAF-receptor antagonists which therefore also may have clinical benefit in the therapy of gastric motility dysfunction and gastroduodenal reflux.

PAF may contribute to microvascular distrubances and ischemic episodes that are associated with many gastrointestinal diseases. Furthermore, the proinflammatory properties of PAF may contribute to inflammatory conditions of the gut, including gastritis and inflammatory bowel disease, and perhaps may interact synergistically with the other inflammatory mediators. The local mucosal release of PAF, along with the release of other proinflammatory mediators such as the leukotrienes may also contribute to the relapse seen in peptic ulceration. Thus, pharmacological agents that interfere with the synthesis or actions of PAF may offer novel therapeutic approaches for the treatment of such gastrointestinal diseases.

Acknowledgments

Figures used in this chapter were adapted from figures prepared by the author in his chapter, PAF and the Gastrointestinal Tract, published in Peter J. Barnes, Clive P. Page, and Peter M. Henson, *Platelet Activating Factor and Human Disease* (Blackwell Scientific Publications, Oxford, 1990).

References

1. P. Bessin *et al., Eur. J. Pharmacol.* **86**, 403 (1983).
2. J.W. Doebber, W.S. Wu, and T.Y. Shen, *Biochem. Biophys. Res. Commun.* **125**, 980 (1984).
3. P. Braquet *et al., Pharmacol. Rev.* **39**, 97 (1988).
4. A.C. Rosam, J.L. Wallace, and B.J.R. Whittle, *Nature (London)* **319**, 54 (1986).
5. J.V. Esplugues and B.J.R. Whittle, *Br. J. Pharmacol.* **93**, 222 (1988) .
6. B.J.R. Whittle *et al., Am. J. Physiol.* **251**, G772 (1986).
7. B.J.R. Whittle, G.L. Kauffman, and J.L. Wallace, in *Advances in Prostaglandin, Thromboxane and Leukotriene Research*, B. Samuelsson, R. Paoletti, and P. Ramwell, Eds. (Raven Press, NY, 1987), pp. 285-292.

8. D.I. Soybel, S.W. Ashley, and L.Y. Cheung, *Am. J. Surg.* **155**, 180 (1988).
9. J.L. Wallace *et al., Gastroenterology* **93**, 765 (1987).
10. A. Brambilia, A. Ghiozi, and A. Giachetti, *Pharmacol. Res. Commun.* **19**, 147 (1987).
11. P. Braquet *et al., Eur. J. Pharmacol.* **150**, 269 (1987).
12. J.L. Wallace and B.J.R. Whittle, *Br. J. Pharmacol.* **89**, 415 (1986).
13. I.W. Lees *et al., Br. J. Pharmacol.* **89**, 504P (1986).
14. K.D. Rainsford, *Pharmacol. Res. Commun.* **18**, Suppl. 209 (1986).
15. J.L. Wallace and B.J.R. Whittle, *Prostaglandins* **32**, 137 (1986b).
16. F. Gonzalez-Crussi and W. Hsueh, *Am. J. Pathol.* **112**, 127 (1983).
17. B.J.R. Whittle *et al., Br. J. Pharmacol.* **92**, 3 (1987).
18. P. Inarrea *et al., Immunopharmacol.* **9**, 45 (1985).
19. T.W. Doebber *et al., Biochem. Biophys. Res. Commun.* **127**, 799 (1985).
20. N.K. Boughton-Smith, I. Hutcheson, and B.J.R. Whittle, *Prostaglandins* **38**, 319 (1989).
21. N.K. Boughton-Smith and B.J.R. Whittle, *Gastroenterology* **94**, A45 (1988).
22. J.L. Wallace, *Can. J. Physiol. Pharmacol.* **66**, 422 (1988).
23. R. Eliakim *et al., Gastroenterology* **95**, 1167 (1988).
24. Z. Terashita *et al., Eur. J. Pharmacol.* **109**, 257 (1985).
25. J. Casals-Stenzel, *Eur. J. Pharmacol.* **135**, 117 (1987).
26. J.L. Wallace and B.J.R. Whittle, *Eur. J. Pharmacol.* **124**, 209 (1986).
27. W. Hsueh, F. Gonzalez-Crussi, and J.L. Arroyave, *Am. J. Pathol.* **122**, 231 (1986).
28. S.R. Findlay *et al., Am. J. Physiol.* **241**, C13O (1981). (1986).
29. N.P. Stimler *et al., Am. J. Pathol.* **105**, 64 (1981).
30. A. Tokumura *et al., Lipids* **18**, 848 (1983).
31. S.B. Hwang *et al., Biochemistry* **22**, 4756 (1983).
32. A. Tokumura, K. Fukuzawa, and H. Tsukatani, *J. Pharm. Pharmacol.* **36**, 210 (1984).
33. J.V. Levy, *Biochem. Biophys. Res. Commun.* 146, 855 (1987).
34. A. Tokumura *et al., Eur. Pharmacol.* **148**, 353 (1988).
35. J.V. Esplugues and B.J.R. Whittle, *Am. J. Physiol.* **256**, G275 (1989).
36. W. Hsueh *et al., Eur. J. Pharmacol.* **123**, 79 (1986).

12

PAF Antagonists:

A Dual Role in Endotoxemia and Sepsis?

Sepsis is a major clinical problem in hospitalized patients. Mortality remains high (30 to 70 percent), with approximately 200 to 400,000 deaths annually in the United States. Endotoxemia-sepsis in animals and man produce both eicosanoids and platelet-activating factor (PAF). The exact role of the eicosanoids or PAF[3] in the pathophysiological mechanisms in endotoxemia and sepsis are unknown. We explored the possibility that PAF receptor antagonists, like the cyclooxygenase inhibitors, would improve the survival and attenuate the eicosanoid release in severe canine and rodent endotoxemia. The results demonstrated that a PAF receptor antagonist, BN 52021, significantly (*P*<0.001) improved the survival in both canine and rodent endotoxemia and surprisingly attenuated eicosanoid release in endotoxemia in both rats and dogs. These findings suggest (1) that an intimate relationship may exist between PAF and the eicosanoids and (2) that the role PAF may have in cellular injury patterns is to amplify the release of other inflammatory mediators.

correspondence
J. Raymond Fletcher
Departments of Surgery
Vanderbilt University and
University of South Alabama
2451 Fillingim Street
721 Mastin Building
Mobile, AL 36617

Mark A. Earnest
A. Gerald DiSimone
Naji Abumrad
James M. Moore
P. Williams
Department of Surgery
Vanderbilt University and

Introduction

Gram-negative sepsis continues to be a substantive problem in all areas of medical practice, accounting for 200 to 400,000 deaths annually in the United States alone. Despite the sophistication of diagnostic methods, aggressive surgical treatment, more potent antibiotics, and highly advanced technology to care for patients, the mortality remains (30 to 70 percent).[1,2] To date the critical mechanisms that are operative in endotoxemia-sepsis are under intense investigation. The present investigation attempts to explore a relationship between a newly recognized inflammatory mediator, PAF, and a well-known group of mediators in sepsis, the eicosanoids.

PAF is released in sepsis in animals[3] and man[4] from activated leukocytes and platelets[5] but could be released from lung, kidney,[6] or endothelium.[7] PAF produces many inflammatory effects: activation of platelets and leukocytes,[8] metabolic dysfunction,[9] and pulmonary injury and plasma extravasation.[10,11] Exogenous PAF is known to release eicosanoids.[12,13] Interestingly, endotoxin administration produces many events that are similar to those observed with exogenous PAF injection.[14,15] Of particular interest to this investigation are reported data demonstrating that exogenous PAF administration stimulates the synthesis and release of eicosanoids (thromboxane, prostacyclin, and leukotrienes).[12,13]

The purpose of this report is to summarize our most recent studies that evaluate the relationship of PAF and the eicosanoids in endotoxemia.

Materials and Methods

Rodent endotoxemia. Male Sprague-Dawley rats (250 to 300g, Charles River Laboratories) were stabilized from two to four days prior to experimentation. They were maintained in a 12-hour light-dark sequence, allowed water and antibiotic-free chow *ad libitum*. On the day of the experiment, animals were placed in individual cages. Each animal was lightly anesthetized with a halothane-oxygen mixture and had an anterior neck incision followed by insertion of catheters (PE50) into the left carotid artery and left jugular vein. Catheters were flushed with heparinized saline, tunneled to the dorsal neck of the rat, and secured. The jugular vein catheter was utilized for injections of saline or pharmacological agents. Rats were allowed to recover from the effects of anesthesia for at least one hour before baseline parameters were measured. At the completion of the experiments, animals were sacrificed.

Canine Endotoxemia. Adult male mongrel dogs (17 to 22kg) were stabilized in large cages for at least seven days prior to study. On the day before the experiments, dogs were anesthetized with Surital, 25mg/kg, I.V. Following intubation and ventilation, a Swan-Ganz catheter was inserted into the pulmonary artery via the right jugular vein utilizing pressure monitoring. From the same neck incision, a silicone catheter was placed in the right carotid artery and advanced into the aorta. The catheters were tunneled subcutaneously to the dorsum of the neck and secured. Catheters were placed in a pocketed vest designed to protect them. Animals were extubated, allowed to recover from anesthesia, and returned to their cages. They were given food and water *ad libitum* until 24 hours before the experiment, then they had only free access to water. Surgical procedures were performed using sterile techniques.

Blood samples were collected in heparinized syringes and immediately placed on ice. The blood was immediately aliquoted for determination of arterial blood gases, hematocrit, hemoglobin, and plasma bicarbonate determination on an acid-base analyzer (Radiometer, ABL-30, Copenhagen, Denmark). *E. coli* endotoxin (055:B5, Difco Laboratories, Detroit, MI) was reconstituted in saline (10 mg/ml) on the day of the experiment. BN 52021 was generously supplied by Dr. Pierre Braquet, Institut Henri Beaufour, Le Plessis Robinson, France, and reconstituted in phosphate buffered saline on the day of the experiment (10mg/ml).

Experimental Design

Rodent endotoxemia. All experiments were performed in an Association for Accreditation for Laboratory Animal Care approved facility and according to the National Institutes of Health (NIH) guidelines for animal use. Rats for hemodynamic and eicosanoid measurements were studied in individual cages. Indwelling catheters were flushed with sterile heparinized saline. There was a 30 minute (–30 to 0 minutes) basal period and a two-hour (0 to 120 minutes) experimental period during which hemodynamic measurements were recorded continuously. These consisted of measurements of heart rate (HR) and mean arterial pressure (MAP). At time 0, the animals received an I.V. bolus injection of endotoxin. In studies where the effect of PAF antagonists were examined, the antagonists were given at specified time points prior to the time 0 minutes.

For the survival studies, male rats were randomized into two groups. Group I (*n*=10) and Group II (*n*=13) received *Escherichia coli* endotoxin (20mg/kg) IP at time 0 minutes. Group I animals received diluent (0.5ml) for the PAF receptor antagonist BN 52021 administered subcutaneously at t =–30 minutes, whereas Group II animals received BN 52021 (25mg/kg) administered subcutaneously at t = –30 minutes. The animals had free access to food and water. Survival was determined at 24, 48, and 72 hours. Permanent survival was determined at 72 hours and is reported in percent.

The animals for hemodynamic and eicosanoid measurements were randomized into two groups following catheter insertion: Group III (*n*=5) received the diluent of BN 52021 at t = –30 min I.V. and then *E. coli* endotoxin (20mg/kg) I.V. at time 0 minutes, whereas Group IV (*n*=5) received BN 52021 (25mg/kg, I.V.) at t = –30 minutes and then the *E. coli* endotoxin (20mg/kg) I.V. at time 0 minutes Blood samples for eicosanoid analysis were collected in heparinized tubes containing indomethacin (100μg), the blood was centrifuged, and the plasma was removed and stored at –80°C until assayed. For each milliliter of blood removed, one milliliter of 0.9 percent saline was returned to the animal via the arterial catheter. Following completion of the experiment, animals were sacrificed.

The selection of doses of endotoxin was determined with dose-response studies in our laboratory. Control studies in animals not given endotoxin were performed in rats with the following groups: (1) anesthesia alone (*n*=4), (2) anesthesia plus incision and ligation of vessels (*n*=5), (3) anesthesia, incision, catheter placement, vehicle administration (saline) for the endotoxin studies (*n*=5); and the diluent for the BN 52021 studies (n=5). None of the perturbations had any significant effect on either the hemodynamic or eicosanoid changes and hence were not included.

Data analysis was accomplished utilizing paired student's t tests for values within the same group and analysis of variance for comparison values between the groups. Differences in survival were determined by the chi-square method. A P value of < 0.05 was considered significant.

Canine endotoxemia. All experiments were performed in an AALAC-approved facility and according to the NIH guidelines for animal use. The dogs were studied in a body support (sling) and allowed to stand during the experiment. Indwelling catheters were flushed with sterile heparinized saline. A 30-minute basal period (–30 to 0 minutes) was followed by a 4-hour experimental period (0 to 240 minutes) during which hemodynamic measurements were recorded continuously. Hemodynamic measurements included HR, MAP, mean pulmonary arterial pressure (PAP), pulmonary capillary wedge pressure (PCWP), central venous pressure (CVP), and cardiac output (CO). Each cardiac output value was determined as the average of two consecutive measurements. Systemic vascular resistance (SVR) and pulmonary vascular resistance (PVR) were calculated from the data [SVR–(MAP–CVP)/CO and PVR–(PAP-PCWP)/CO)] and expressed in arbitrary units. At time zero, each animal received a bolus injection of endotoxin (1mg/kg, I.V.)

Blood samples were obtained at time zero, then at 2, 60, 120, and 240 minutes after endotoxin administration. Hemodynamic parameters were determined at the same time as the blood sampling. Animals were randomized into two groups: (1) Group I: endotoxin-vehicle (n=10) received an LD_{100} dose of *E. coli* endotoxin (1mg/kg) I.V. at time zero, and (2) Group II: endotoxin-BN 52021 (n=10) received BN 52021 (5mg/kg) I.V. 30 minutes before and 240 minutes after *E. coli* endotoxin administration. All animals received initial volumes of saline (0.9 percent, w/v) to standardize the PCWP to 6 torr before the experiment. During the four-hour experiment, 400ml of balanced salt solution were administered, and at the end of four hours an additional 400ml was given over 30 minutes. The catheters were removed, and the animals were returned to their individual cages. They were given free access to food and water. Survival was determined at 24, 48, and 72 hours. Only five animals in each group were used to determine survival, as dictated by our animal-use committee.

For hemodynamic measurements of both rats and dogs, the catheters were connected to a Gould Brush physiograph (RS2300) via a PE23 transducer (Statham, Oxnard, CA). Cardiac outputs of the dogs were calculated with an Edwards computer (Model 9320) using a thermodilution method. Levels of thromboxane (Tx) and prostaglandin E_2 (PGE_2) were measured in plasma specimens collected at baseline, then at 5, 30, 60, and 120 minutes after endotoxin administration in coincident with hemodynamic measurements. Analyses were performed by radioimmunoassay. Cross-reactivity of the antibodies is less than 3 percent with other eicosanoid metabolites. Radiolabeled TxB_2, 6-keto-PGF_1, and PGE_2 were obtained from New England Nuclear (Boston, MA). Authentic eicosanoids TxB_2, PGI_2, and PGE_2 were generously supplied by Dr. John Pike (Upjohn Co., Kalamazoo, MI). *Escherichia coli* endotoxin (055:B5, Difco Laboratories, Detroit, MI) was reconstituted in saline (10mg/ml) on the day of the experiment. The PAF receptor antagonist, BN52021, was supplied by Dr. Pierre Braquet, Institut Henri Beaufour, Le Plessis Robinson, France. BN52021 was diluted in a phosphate solution (pH 7.3) on the day of the experiment (10mg/ml).

The survival data were analyzed using a chi-square test. For the other parameters, the two treatments were compared using a repeated-measures ANOVA. Within each treatment group, the baseline values were compared to each time point using a paired student's t test. A probability of $P < 0.05$ was considered significant.

Results

Rodent endotoxemia. The endotoxin alone significantly ($P < 0.05$) decreased MAP from 112 ± 7 torr (baseline) to 74 ± 7 (four hours), whereas in the PAF receptor antagonist treated group, there was no significant decrease in the MAP at the four hour (100 ± 4 torr) time point. Of interest is the observation that with the PAF receptor antagonist, there was no prevention of the early significant decrease in MAP (five minutes) following endotoxin administration (88 ± 3 versus 90 ± 8 torr).

Endotoxin alone significantly stimulated TxB_2 synthesis at all time points measured from baseline: (424 ± 56pg/ml) to peak at 60 minutes (1561 ± 146) ($P < 0.001$), whereas with PAF receptor antagonism the TxB_2 values were only different from the baseline (368 ± 88pg/ml) at the 120-minute time point (734 ± 93pg/ml, $P <$ 0.05). PGE_2 values were similar to those for TxB_2, and PAF antagonism significantly attenuated its release at all time points.

PAF receptor antagonism (BN 52021) improved the permanent survival (11 of 13, 85 percent) when compared with survival in the group that received endotoxin alone (2 of 10, 20 percent), $P < 0.01$.

Canine endotoxemia. Endotoxin alone significantly ($P < 0.05$) increased the heart rate:baseline (116 ± 6 bpm) versus that at 240 minutes (149 ± 8 bpm); decreased the MAP:baseline (123 ± 5 torr) versus one to two minutes (54 ± 5 torr) ($P < 0.05$); decreased the CO:baseline (3.09 ± .31 liters per minute) versus one to two minutes (1.26 ± .2 liters per minute) ($P < 0.\ 05$); increased the PVR:baseline (3.2 ± .6 units) versus one to two minutes (13 ± 3 units) ($P < 0.0$). With the PAF receptor antagonism (BN 52021) treatment, there was attenuation of the increase in heart rate and the increase in PVR, but no effect on the dramatic decreases in CO or MAP.

For all the other parameters investigated (hematocrit, pH, partial CO_2 pressure, partial oxygen pressure, bicarbonate radical, white blood cell count, platelet count), pretreatment with BN 52021 did not prevent the metabolic acidosis, hypocarbia, and the alterations in WBC counts or thrombocytopenia.

Endotoxin alone significantly ($P < 0.05$) enhanced release of all eicosanoids (TxB_2, PGE_2, 6-keto-PGF_1): TxB_2 baseline (0. 26 ± .04ng/ml) versus 240 minutes (2.64 ± .96) ($P < 0.05$); PGE_2 baseline (0.20 ± .10ng/ml) versus 240 minutes (0.79 ± .35); 6-keto-PAF_1 baseline (0.140 ± .05ng/ml) versus peak at 60 minutes (2.07 ± .84) ($P < 0.05$).

Administration of PAF receptor antagonist attenuated the increases in TxB_2 at 60, 120, and 240 minutes but had minimal effect on PGE_2 and 6-keto values.

There was a significant ($P<0.01$) increase in permanent survival with PAF receptor antagonist pretreatment [100 percent (5 of 5) versus 0 percent (0 of 5)] in canine endotoxemia.

Discussion

The results of these data support the notion that PAF actions have a substantive role in the pathophysiology of severe endotoxemia in two different animal models. PAF receptor antagonism improves survival and dramatically alters eicosanoid synthesis and release. Surprisingly, PAF receptor antagonism with BN 52021 had minimal to no effect on the early dramatic hemodynamic events observed in rodents or canines following endotoxin injection.

Improvement in survival and other pathophysiological parameters in endotoxemia with PAF receptor antagonists has been reported by others.[16-19] Our present finding with respect to the absence of significant alterations in the hemodynamic events by PAF receptor antagonists are in contrast to other citations.[18,19] These differences may be related to species variation, the concentration and type of endotoxin utilized, methods of evaluation, and differences in the models developed. We are unaware of reported data that demonstrate PAF receptor antagonism improves survival in a resuscitated canine model.

Enhanced eicosanoid synthesis and release in endotoxemia-sepsis has been well documented in humans as well as in animal models of sepsis. The role of PAF is less certain; however, exogenous PAF is known to release eicosanoids. Because the precursors of PAF and the eicosanoids may be similar,[20,21] we wished to explore this relationship in endotoxemia. In the rat, PAF receptor antagonism appeared to be more effective in attenuating eicosanoid release compared with the canine model. Of interest in this regard is that the rat platelet is known to lack a PAF receptor. PAF receptor antagonism more directly influenced TxB_2 release when compared with PGE_2 release following endotoxin injection in canines. The value of these observations is unclear, and additional studies are needed.

That an intimate relationship exists between PAF and the eicosanoids is supported by recent reports. Acute circulatory collapse in dogs induced by exogenous PAF is associated with increased plasma values of TxB_2 and 6-keto-PGF_1.[22] PAF effects on coronary blood flow have been related to TxB_2 presence.[23] Pulmonary hypertension and increased vascular permeability in the lung induced by exogenous PAF may be caused by a cyclooxygenase dependent and cyclooxygenase independent mechanism.[24] The injection of PAF in an isolated gastric perfusion model demonstrated an increase in eicosanoid release, and PAF receptor antagonism (BN 52021) attenuated the vasoconstriction and the eicosanoid release.[25] Several biochemical studies suggest a relationship between PAF and the eicosanoids.[26]

Albert and Snyder[26] indicated that a phospholipid, 1-alkyl-2-acyl-GPC, exists in the cell membrane, and others suggest this phospholipid is a precursor of both PAF and arachidonic acid metabolites in human platelets,[27] rat and rabbit alveolar macrophage,[28] and peritoneal macrophage from guinea pig and rabbit neutrophils.[29] Chilton *et al.*[30] demonstrated *in vitro* the release of arachidonate from cellular phospholipids in cytochalasin B-treated rabbit polymorphonuclear leukocytes. The phospholipid sources of the arachidonate were phosphatidylinositol and phosphatidylcholine (~50 percent from each). Interestingly, the PAF-stimulated production of arachidonate metabolites and the degranulation response was blocked by eicosatetraenoic acids and nordihydroquaiaretic acid (lipooxygenase inhibitors). These investigators[30] implied that the bioactions of PAF on polymorphonuclear leukocytes may be mediated, in part, by the release of arachidonic acid, resulting in production of mono- and dihydroxy-eicosatetraenoic acids [5(S)-hydroxy-6,8 eicosatetraenoic acid; leukotriene B_4]. In additional experiments, they[31,32] demonstrated that the metabolic fate of PAF included an incorporation of arachidonate in rabbit and human polymorphonuclear neutrophils (PMN); in human PMN, 75 to 80 percent of the added PAF is reacylated with arachidonate. The most recent report[32] indicates that PAF precursors (1-0-alkyl-2-arachidonyl-GPC) are a common source of both PAF and arachidonate through the action of phospholipase A_2.

In support of the *in vitro* findings of Chilton *et al.*,[30] Haroldsen *et al.*[34] demonstrated, in the intact lung, that 20 to 23 percent of the administered PAF resulted in 1-0-hexadecyl-2-arachidonyl-GPC. These data support the hypothesis that a relationship between PAF and arachidonate metabolism exist in the intact organ.

The preceding information suggests that exogenous PAF elicits the production of both lipooxygenase and cyclooxygenase products, presumably from arachidonic acid. In some instances, the effects of PAF are dose-dependent, and in others of the effects of PAF are mediated via the effects of the eicosanoid metabolites. The fundamental cellular interrelationships are less certain. When cellular injury occurs (by many mechanisms), phospholipase A_2 is activated;[34] therefore, the potential for simultaneous release of arachidonate (fatty acids) and PAF is present. The studies by Chilton[30] and Haroldsen[34] *vide supra* indicate that PAF precursors may be responsible for a portion of the arachidonic acid released, which suggests an intimate relationship does exist between PAF and the eicosanoids.

The finding in this study that PAF receptor antagonism attenuated the eicosanoid release following endotoxin administration was unexpected. That PAF receptor antagonism significantly improved the survival in these severe models of endotoxemia supports the notion that PAF may have a substantive role in the cellular events relating to the pathophysiology of endotoxemia. Additional studies are needed to further elucidate the interrelationships of PAF to the eicosanoids and further define the signal mechanisms that are critical in the early events in endotoxemia-sepsis.

References

1. R.C. Bone *et al., New Engl. J. Med.* **317**, 653 (1987).
2. The Veterans Administration Systemic Sepsis Cooperative Study Group, *New Engl. J. Med.* **317**, 659 (1987).

3. P. Inarrea *et al., Immunopharmacology* **9,** 45 (1985).
4. F. Bussolino *et al., Thromb. Res.* **48,** 619 (1987).
5. J. Benveniste *et al., Thromb. Res.* **25,** 375 (1982).
6. W. Renooij and R. Snyder, *Biochem. Biophys. Acta* **663,** 545 (1981).
7. G. Camussi *et al., J. Immunol.* **131,** 2397 (1983).
8. M. Sanchez-Crespo *et al., Pharmacol. Res. Commun.* **18,** 181 (1986).
9. P. Bessin *et al., Eur. J. Pharmacol.* **86,** 403 (1983).
10. M. Mojarad, Y. Hamasaki, and S.E. Said, *Bull. Eur. Physiopathol. Respir.* **19,** 253.
11. C.G. Caillar *et al., Agents Actions* **12,** 725 (1982).
12. A. Otsuka *et al., Prostaglandins Leukotrienes Med.* **19,** 25 (1985).
13. A. Dembinska-Kiec *et al., Prostaglandins* **37,** 69 (1989).
14. J.R. Fletcher anc P.W. Ramwell, *J. Surg. Res.* **24,** 154 (1978).
15. J.R. Fletcher and P.W. Ramwell, *Br. J. Pharmacol.* **64,** 185 (1978).
16. S. Adnot *et al., Prostaglandins* **32,** 791 (1986).
17. J. Casals-Stenzel, *Eur. J. Pharmacol.* **135,** 117 (1987).
18. T. Doebber *et al., Biochem. Biophys. Res. Commun.* **3,** 799 (1985).
19. D. Handley *et al., Immunopharmacology* **13,** 125 (1987).
20. M. Chignard, D. Delantier, and J. Benveniste, *Biochem. Biophys. Res. Commun.* **124,** 637 (1984).
21. F.H. Chilton *et al., J. Biol. Chem.* **259,** 12014 (1984).
22. P. Bessin *et al., Eur. J. Pharmacol.* **86,** 403 (1986).
23. D. Erza *et al., Adv. Prostaglandin Thromboxane Leukotriene Res.* **13,** 14 (1985).
24. K.E. Burhop *et al., Am. Rev. Resp. Dis.* **134,** 548 (1986).
25. A. Dembina-Kiec *et al., Prostaglandins* **37,** 69 (1989).
26. D.H. Albert and F. Snyder, *J. Biol. Chem.* **258,** 97 (1983).
27. V. Natarajan *et al., Throm. Res.* **30,** 119 (1983).
28. H.W. Mueller, J.T. O'Flaherty, and R.L. Wykle, *Lipids* **17,** 72 (1982).
29. T. Sugiura *et al., Biochem. Biophys. Acta* **712,** 515 (1982).
30. F.H. Chilton *et al., J. Biol. Chem.* **257,** 5402 (1982).
31. Ibid., **258,** 6357 (1983).
32. Ibid., **258,** 7268 (1983).
33. Ibid., **259,** 12014 (1984).
34. P.E. Haroldsen *et al., J. Clin. Invest.* **79,** 1860 (1987).

13

PAF And TNFα Relationship in Septic Shock

Reuven Rabinovici
Tian Li Yue
correspondence
Giora Feuerstein
Department of Pharmacology
SmithKline Beecham
Pharmaceuticals
King of Prussia, PA
19406-0939

Septic shock is characterized by episodic bacteremia associated with circulatory collapse and multi-system failure. Both platelet-activating factor (PAF) and tumor necrosis factor-α/cachectin (TNFα) have emerged as important mediators of the septic shock syndrome for the following reasons: (1) similarities between endotoxin and PAF or TNFα biological effects; (2) elevation of circulating PAF and TNFα levels during endotoxemia; and (3) protective effects of PAF antagonists and anti-TNFα antibodies in the septic state. Therefore, it is of interest to unravel the plausible interactions that link PAF and TNFα in this pathophysiological state. This chapter presents an overview of the role of PAF and TNFα in septic shock and compiles the available evidence on the relationship between these two mediators.

PAF Synthesis and Metabolism

PAF is an alkyl-acyl-phosphorylcholine with the chemical structure of 1-O-alkyl-2-acetyl-*sn*-glycero-3-phosphocholine.[1,2] The hormone is produced by macrophages, neutrophils, natural killer lymphocytes, platelets, endothelial cells,[3-5] and neuronal cells.[6] The predominant organs capable of PAF synthesis are the kidney,[7] heart,[8,9] brain,[10,11] lung,[12] and intestine.[13] It is beyond the scope of this review to discuss the extensive data published on the biosynthesis of PAF, a subject that has been thoroughly reviewed in recent literature.[14,15] In brief, PAF can be formed by two distinct pathways: (1) the *de novo* pathway, which maintains the physiological levels of PAF in the circulation and tissues through its production from 1-alkyl-2-1yso-*sn*-glycero-3-P via alkylacetylglycerol. The key enzyme in this metabolic route, alkylacetylglycerol:CDP choline choline-phosphotransferase, is not stimulated or modified by endotoxin and inflammatory agents.[16] (2) the *remodeling* pathway, in which PAF is synthesized from alkyl-acyl-glycerophosphocholines via lysoPAF. This pathway is probably responsible for the modulation of PAF responses to various pathophysiological states because the major enzyme, 1-O-alkyl-2(R)-lyso- glycero-3-phosphocholine, can be activated by inflammatory agents such as thrombin,[17] interleukin 1, and TNFα.[18] The remodeling pathway also produces arachidonic acid, the precursor of the prostaglandin mediators, which in turn regulates this pathway of PAF synthesis, presumably by stimulation of phospholipase A2 activity.[19] PAF is inactivated in the cell[20] and in the plasma[21,22] by cytosolic or plasma acetylhydrolase[20] to form the inactive lysoPAF.

TNFα Synthesis and Metabolism

TNFα is a 17,000-dalton polypeptide (157 residues in human TNFα) arranged in dimeric, trimeric, or pentameric forms depending on species and method of isolation[23-26] (see Figure 1). TNFα is secreted *in vivo* as a propeptide (extra 76 amino acids in the human form) primarily by macrophages. It is interesting that all types of macrophages thus far tested, including those from pulmonary, hepatic, splenic, peritoneal, and bone marrow origin,[27] were found to produce TNFα. Several other types of cells were reported to produce small amounts of TNFα in response to inducing stimuli. For example, T lymphocytes, induced by the calcium ionophore A23187 in conjunction with phorbol myristate acetate [but not by lipopolysaccharide (LPS)], synthesized TNFα mRNA and protein.[28] Smooth muscle cells were also reported to produce TNFα mRNA,[29] whereas TNFα levels were

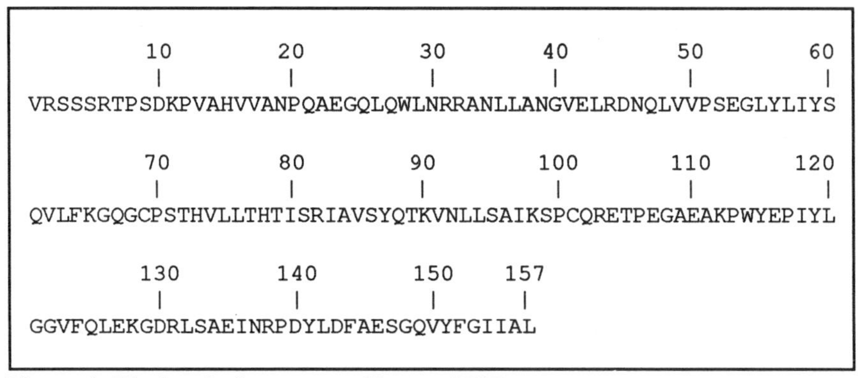

Figure 1. Biochemical structure of human TNFα.

undetectable. Stimulation of mouse astrocytes and microglia in enriched culture by endotoxin resulted in the production of TNFα.[30] Although the signals and processes associated with the expression of TNFα genes are still obscure, endotoxin was shown to accelerate the production of TNFα mRNA, whereas glucocorticosteroids strongly inhibit TNFα gene transcription and the mobilization of the mRNA once it has been formed.[27,28]

PAF Role in Septic Shock

The evolving concept in the pathophysiology of septic shock suggests that endotoxin induces its detrimental effects indirectly through multiple mediators. Since the elucidation of its chemical structure in 1979,[1,2] PAF was implicated as one of the major mediators of septic shock. Despite the limitations of the techniques employed for PAF extraction and assay and the short circulating half-life of PAF (hydrolyzed rapidly in the plasma), several studies were able to provide biochemical evidence for the production of PAF during endotoxemia. For example, PAF was observed to be released in the plasma of septic children.[31] Also, Chang *et al.*[32] reported a threefold elevation in plasma PAF and a tenfold elevation of PAF levels in whole lung homogenate 20 minutes following intraperitoneal injection of Salmonella enteritidis endotoxin. It is interesting that these levels persisted for at least two hours, possibly affecting distant organs where no PAF production took place. Although renal PAF levels were also increased following endotoxin injection, they lagged both in time and magnitude. Elevated circulating levels of PAF (measured by a rabbit platelet aggregation bioassay after partial purification) were also reported by Doebber *et al.*[33] in rats exposed to *Escherichia coli* LPS. In this latter study, PAF plasma levels were detectable only at ten minutes following the injection of endotoxin (but not at 1.5 minutes), thus correlating with the LPS-induced hypotension, which commenced three minutes post injection. Because PAF decreases systemic arterial pressure immediately, it is conceivable that the delayed elevation of circulating PAF levels reflected the time required for PAF synthesis and secretion.

Additional evidence for the production of PAF during endotoxemia was presented by Innarea *et al.*,[34] who demonstrated enhanced release of PAF (measured by the [^{3}H]serotonin release from washed rabbit platelets) from spleen lymphocytes and peritoneal cells taken from rats subjected to *E. coli* peritonitis. Significant amounts of PAF in the peritoneal protein-rich exudate could be obtained at 30 minutes after the inoculation of *E. coli*, whereas PAF in peritoneal and splenic leukocytes was elevated only at two hours following the injection of endotoxin. The endotoxin-induced elevation of PAF levels in the peritoneal and splenic cells occurred before the PAF-induced blood volume depletion, suggesting that hemoconcentration and volume contraction are secondary to PAF production. The relative contribution of spleen lymphocytes and peritoneal cells to the overall production of PAF in endotoxemia is still obscure, and the investigation of PAF sources during endotoxemia has not as yet been properly addressed. Several *in vivo* and *in vitro* reports have suggested that PAF is released from injured pulmonary endothelial cells. For example, rabbit pulmonary endothelial cells injured by goat anti-rabbit lung angiotensin converting enzyme gamma globulin[35,36] released PAF into the circulation. Also, elevated PAF levels were found in whole lung homogenate of rats challenged with endotoxin.[32] Taken together, these data suggest that endothelial pulmonary cells might play a major role in PAF production during septic shock.

The detection of specific binding sites for PAF on platelets, leukocytes, and lung tissue[37] and the subsequent development of various PAF antagonists[38] provided useful tools to further implicate PAF as an important mediator of endotoxin shock. For instance, several structurally different PAF antagonists such as ONO-6240,[39] SRI 63-441,[40] SRI 63-072,[41] kadsurenone,[33] WEB 2086,[42] CV 3988,[32,43] BN 52021,[44] and L-652,731[45] were shown to protect against endotoxin-induced hypotension in several species. Also, the PAF antagonist ONO-6240 abolished the *E. coli*- induced decrease in cardiac output observed in conscious sheep given endotoxin,[39] and the PAF antagonist CV-3988 protected against endotoxin-induced low cardiac output and increased peripheral resistance. Blood flow to essential organs such as heart, kidney, and the splanchnic bed was also substantially improved among the nontreated septic group.[46] Moreover, other studies, using a large variety of PAF antagonists, have shown protection or reversal of endotoxin-induced leukopenia, thrombocytopenia, and hemoconcentration. For example, FR-900452, tested in anesthetized rabbits,[47] and the specific PAF antagonist BN 52021, injected to anesthetized guinea pigs,[44] were shown to attenuate endotoxin-induced thrombocytopenia. Moreover, ONO-6240 partially prevented the leukopenia induced by endotoxin in conscious sheep.[39] Most important, pretreatment with PAF antagonists such as CV 3988[32,43] BN 52021,[48,49] and WEB 2086[42] improved survival of rats and dogs exposed to various endotoxins.

TNFα Role in Septic Shock

Like PAF, TNFα was implicated as a major mediator of the septic shock syndrome. The observations that suggested TNFα is as an endotoxin-induced systemic mediator included: (1) plasma levels of TNFα were found to be elevated in patients with meningococcal meningitis, septicemia, or both.[50] The detection of TNFα seemed to

Table 1
Serum TNFα Levels in Endotoxic Shock

Endotoxin	Dose/Mode	Species	TNFα	Reference
n.p.	400μg/animal IP	Mouse	6.2×10^3U/ml at 60 minutes	79
E. coli	0.1 mg/kg IV	Rat	1.6×10^5U/ml at 90–120 minutes	62
E. coli	5mg/kg IV	Rat	3.5×10^5U/ml at 60–90 minutes	80
E. coli	1mg/kg IV	Rat	9.5×10^3U/ml at 90 minutes	81
S. minessota	10μg/animal IV	Rabbit	2.5×10^3U/ml at 45–100 minutes	55
E. coli	30μg/animal IV	Rabbit	9.5×10^3U/ml at 120 minutes	82
E. coli	1.2×10^{11}bac/kg IA	Baboon	1.7×10^3U/ml at 120 minutes	54
E. coli	10^{11}–10^{12}bac/kg IV	Baboon	810U/ml at 90 minutes	83
E. coli	4ng/kg IV	Human	10U/ml at 120 minutes	84
E. coli	20μg/kg IV	Human	14U/ml at 90 minutes	83

Abbreviations: IV = intravenous; IA = intraarterial; IP = intraperitoneal; bac = live bacteria.; n.p. = not published.

correlate with the degree of disease because TNFα levels were elevated in 10 of the 11 patients who died but were elevated in only 8 of 68 survivors; and (2) systemic administration of LPS elevated circulating levels of TNFα in all species studied (Table 1). TNFα response is monophasic, with a peak at 90 to 120 minutes and a

Table 2
Biological Responses to Administration of Endotoxin, PAF, and TNFα

Biological Response	Endotoxin	PAF	TNFα
Blood pressure	↓	↓	↓
Heart rate	↑	↑	↑
Cardiac output	↑ or ↓*	↓	ϕ or ↓
Peripheral resistance	↓ or ↑*	↓	ϕ or ↑
Platelets	↓	↓	↓
Leukocytes	↓	ϕ	↓
Hematocrit	↑	↑	↑
TXB_2	↑	↑	?
6-keto-$PGF_{1\alpha}$	↑	ϕ	?
Epinephrine	↑	↑	↑
Norepinephrine	↑	↑	↑
Cortisol	↑	↑	↑

Symbols: ↑ increase; ↓ decrease; ϕ no effect; ? unknown; * dependent on phase of septic shock.

magnitude varying with dose of LPS and species; (3) once induced by endotoxin, monocytes and macrophages released copious amounts of TNFα;[51,52] (4) passive immunization against TNFα in several species protected from lethal effects of endotoxin. For example, mice pretreated with a highly specific polyclonal rabbit antiserum directed against murine TNFα were protected in a dose-dependent manner against endotoxin-induced lethality.[53] Also, passive immunization with monoclonal anti-TNFα $F(ab')_2$ fragments protected baboons against shock induced by an LD_{100} dose of *E. coli*.[54] In rabbits, prior infusion of polyclonal anti-TNFα antibodies (prepared by immunizing goats with human recombinant TNFα) neutralized the endotoxin-induced hypotension, fibrin deposition, and lethality;[55] (5) endotoxin-resistant mice (C3H/HeJ) did not produce large amounts of TNFα following exposure to LPS;[56] and (6) the administration of TNFα closely mimics the clinical presentation of endotoxic shock (Table 2). For example, TNFα produced hypotension and tachycardia in anesthetized and conscious rats,[57,58] conscious guinea pigs,[59] conscious rabbits,[60] and anesthetized dogs.[61] Also, TNFα induced hemoconcentration, leukopenia, and thrombocytopenia in all species studied.[57,58,60,62-64] Moreover, Nathanson *et al.*[65] showed that in dogs given a single intravenous bolus of recombinant TNFα, dilation of the left ventricle and decrease of the ejection fraction occurred, which are phenomena also seen in septic shock.

PAF/TNFα Interaction in Septic Shock

Several *in vitro* studies in cell systems have suggested a possible link between PAF and TNFα. Cammusi *et al.*[18] observed that murine (1 to 10ng/ml) and human (10 to 50ng/ml) TNFα stimulated PAF production by cultured rat peritoneal macrophages, rat polymorphonuclear neutrophils, and human vascular endothelial cells cultured from the umbilical cord vein. A possible mechanism of this interaction could be the activation of 1-O-alkyl-2(R)-lyso-glycero-3-phosphocholine:acetyl Coenzyme A acetyltransferase,[16,18,66,67] a key enzyme in the biosynthesis of PAF, by TNFα. It is interesting that PAF (5 to 50ng/ml for 1 to 4 hours) did not induce appreciable release of TNFα into the culture medium of rat peritoneal macrophages. In contrast, Hayashi *et al.*[68] reported that PAF (10^{-5}M) can release TNFα from guinea pig peritoneal macrophages with maximal activity at eight hours. Also, Bonavida *et al.*,[69,70] using a sensitive radioimmunoassay, demonstrated that PAF facilitates in a dose-dependent manner the production of TNFα by peripheral blood-derived monocytes; TNFα activity was present initially but declined after 24 hours. Additional evidence that PAF induces TNFα production was provided by Rola-Pleszczynski *et al.*[71] who used cell cultures of human and rodent lymphocytes and macrophages. *In vivo* evidence in support of TNFα-induced PAF production was obtained in studies where TNFα (0.5 to 1.0mg/kg) increased PAF synthesis by intestinal tissue.[72]

Other studies have reported interactions between PAF- and TNFα-induced endothelial cell injury. For example, in guinea pigs superfused with TNFα, a subsequent injection of a low-dose PAF into the mesenteric vasculature resulted in increased thrombosis. This synergistic effect was inhibited by pretreatment with the PAF antagonist BN 52021 or anti-TNFα antibodies (Bourgain and Braquet, in preparation). PAF antagonists also attenuated priming of polymorphonuclear neutrophils by TNFα, including production of oxygen radicals and leukotrienes.[73] Priming relationships between PAF and TNFα have been recorded in two other models: (1) pretreatment of rats with TNFα amplified the extravasation induced by a low and nonactive dose of PAF[74] and (2) the combined administration of nonactive doses of PAF and TNFα caused a significant loss of glycosaminoglycans in cartilage tissue.[75]

Unfortunately, only a handful of studies are available in which PAF/TNFα interactions were evaluated in relationship to endotoxin challenge. Recently, pretreatment or posttreatment with the new PAF antagonist RP 55778 was reported to inhibit TNFα production by isolated murine macrophages exposed to *Salmonella enteritidis* endotoxin.[76] Also, high concentrations of RP 55778 added simultaneously to endotoxin completely inhibited the TNFα mRNA signal. Another line of support to the possible linkage of PAF and TNFα in septic states was provided by Heuer and Weber[77] who have demonstrated that pretreatment of mice with either *S. typhosa* endotoxin or TNFα resulted in a synergistic increase in PAF-induced mortality. It should be noted that when administered alone at similar doses, both TNFα and endotoxin did not effect mortality.

To date, there are no published *in vivo* studies on the interrelations between PAF and TNFα in endotoxemic or septic shock. Nevertheless, some indirect information was obtained from a recent report by Myers *et al.*[78] who observed no beneficial effect of pretreatment or posttreatment with the PAF antagonist BN 52021 on the lethality of combined low dose endotoxin/recombinant human TNFα in mice. It is interesting that these low doses were inactive when endotoxin or recombinant human TNFα were injected alone. In addition, preliminary studies from our laboratory support the possibility that PAF and TNFα interact with each other in endotoxic shock. For example, pretreatment with the new PAF antagonist BN 50739 attenuated endotoxin-induced elevation of serum TNFα (Figure 2) and recombinant human TNFα-induced lethality in rats. In contrast, when administered before or after the injection of endotoxin to rabbits, BN

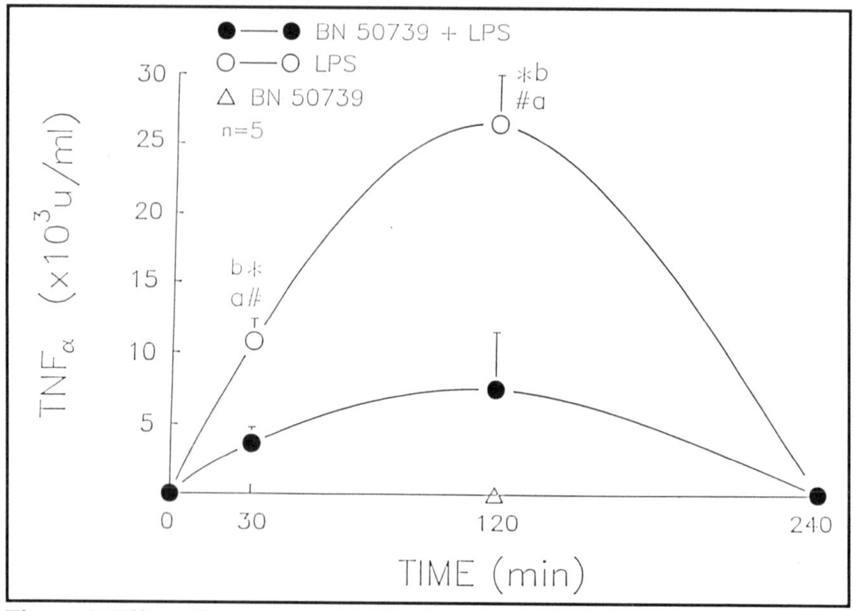

Figure 2. Effect of pretreatment (–30 minutes) with BN 50739 (10mg/kg, IP) on TNFα response to endotoxin (14.4mg/kg, IV) in the rat. LPS = lipopolysaccharide; a = versus basal value; b = versus other group; * = *P*<0.05, # = *P*<0.01.

50739 did not effect the increase in serum TNFα activity induced by endotoxin (Figure 3). Taken together, these studies suggest that in septic shock, PAF and TNFα display reciprocal relationships that lead to deviation amplification of the pathophysiological consequences of endotoxin. Furthermore, because the plasma levels of PAF peak as early as 20 minutes after the administration of LPS,[32] whereas endotoxin-induced elevation of plasma TNFα peaks later at 90 to 120 minutes following the endotoxic insult,[62] it is conceivable that there is a positive feedback loop where PAF serves as a "trigger" for the production of TNFα which, in turn, acts later to release PAF.

Conclusion

Both PAF and TNFα have been implicated as mediators of septic shock. Preliminary *in vitro* and *in vivo* studies indicate that these two

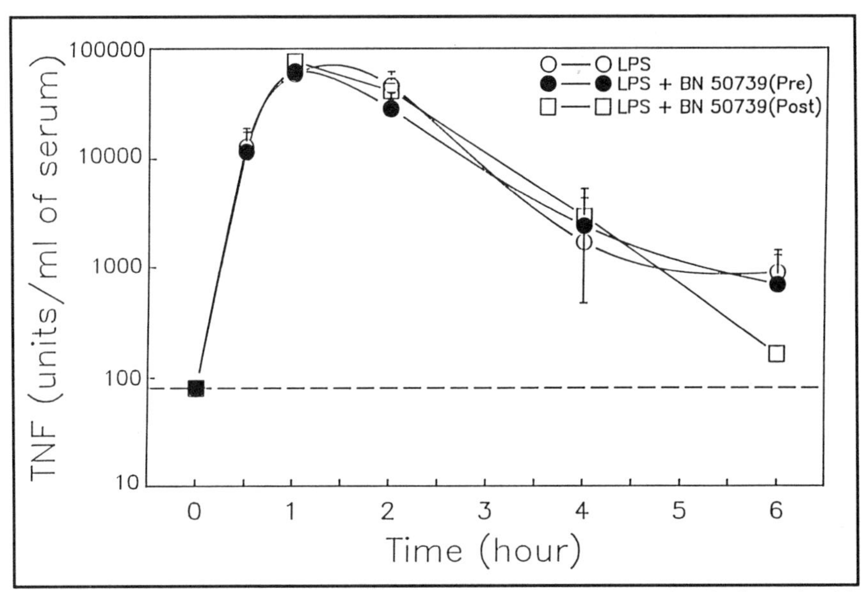

Figure 3. Effect of pretreatment (–30 minutes) or posttreatment (+15 minutes) with BN 50739 (10mg/kg, IP) on TNFα response to endotoxin (50μg/kg, IV) in the rabbit. $n = 9$; LPS = lipopolysaccharide.

mediators are interrelated, possibly through positive feedback loops. However, because the septic shock syndrome involves multiple systems and is associated with severe hemodynamic, hematological, biochemical, and metabolic consequences, and because multiple mediators (for example, leukotriens, catecholamines, prostaglandins, glucocorticosteroids, and so forth) are produced during the septic condition, it is conceivable to assume that many complex interactions of various degree of importance occur in each time point during the stormy events of endotoxic shock. The relative contribution of PAF/TNFα interactions to the overall end result of systemic exposure to endotoxin is still unknown and should await further investigations.

References

1. C.A. Demopoulos, R.N. Pinckard, and D.J. Hanahan, *J. Biol. Chem.* **254**, 9355 (1979).
2. J. Benveniste *et al., C. R. Acad. Sci. Ser. D.* **289**, 1037 (1979).
3. Camussi G, Kidney Int. **29**, 469 (1986).

4. R.N. Pinckard, L.M. McManus, and D.J. Hanahan, *Adv. Inflammation Res.* **4**, 144 (1982).
5. F. Snyder, *Med. Res. Rev.* **5**, 107 (1985).
6. T.L.Yue, P.G. Lysko, and G. Feuerstein, *Neurochem.*, in press.
7. E. Pirotzky *et al., Kidney Int.* **25**, 404 (1984).
8. G. Camussi *et al., Transplantation* **44**, 113 (1987).
9. R. Levi *et al., Cir. Res.* **54**, 117 (1984).
10. K.F. Clay and R.C. Baker, in *Proceedings of the 33rd Annual Conference on Mass Spectrometry and Allied Topics*, (San Diego, CA, 1985), p. 700.
11. R. Kumar *et al., Biochim. Biophys. Acta* **963**, 375 (1988).
12. M.F. Fitzpatrick, S. Moncada, and L. Parente, *Br. J. Pharmacol.* **88**, 149 (1986).
13. J. Filep *et al., Biochem. Biophys. Res. Commun.* **158**, 353 (1989).
14. P. Braquet *et al, Pharmacol. Rev.* **39**, 97 (1987).
15. F. Snyder, *Proc. Soc. Exp. Biol. Med.* **190**, 125 (1989).
16. F. Bussolino *et al., Biochim. Biophys. Acta* **927**, 43 (1987).
17. S.M. Prescott, G.A. Zimmerman, and T.M. McIntyre, *Proc. Natl. Acad. Sci. USA* **81**, 3934 (1984).
18. G. Cammusi *et al., J. Exp. Med.* ***166**, 1390 (1987).*
19. *M.M. Billah, R.W.* Bryant, and M.I. Siegel, *J. Biol. Chem.* **260**, 6899 (1985).
20. M.L. Blank *et al., J. Biol. Chem.* **256**, 175 (1981).
21. M.L. Blank *et al., Biochem. Biophys. Res. Commun.* **113**, 666 (1983).
22. R.S. Farr *et al., Fed. Proc.* **42**, 3120 (1983).
23. B. Beutler *et al., J. Exp. Med.* **161**, 984 (1985).
24. B. Beutler *et al., Nature (London)* **316**, 552 (1985).
25. B.B. Aggarwal *et al., J. Biol. Chem.* **260**, 2345 (1985).
26. S. Abe *et al., FEBS Lett.* **180**, 203 (1985).
27. T. Decker, M.L. Lohmann-Matthes, and G.E. Gifford, *J. Immunol.* **138**, 957 (1987).
28. M.C. Cuturi *et al., J. Exp. Med.* **165**, 1581 (1987).
29. P. Libby, S.J.C. Warner, and C.B. Galin, *Clin. Res.* **35**, 297A (1987).
30. M. Sawada *et al., Brain Res.* **491**, 394 (1989).
31. F. Bussolino *et al., Thromb. Res.* **48**, 619 (1987).
32. S.W. Chang *et al., J. Clin. Invest.* **79**, 1498 (1987).
33. T. Doebber *et al., Biochem. Biophys. Res. Commun.* **127**, 799 (1985).
34. P. Inarrea *et al., Immunopharmacology* **9**, 45 (1985).
35. G. Camussi *et al., J. Immunol.* **131**, 1802 (1983).
36. Ibid., 2397.
37. F.H. Valone, in *Platelet Activating Factorand Related Lipid Mediators,* F. Snyder, Ed. (Plenum Press, New York, 1987), pp. 137-151.
38. P. Braquet *et al., Pharmacol. Rev.* **39**, 97 (1987).
39. T. Toyofuko *et al., Prostaglandins* **31**, 271 (1986).
40. D.A. Handley *et al, Immunopharmacology* **13**, 125 (1987).
41. D.A. Handley, R.C. Anderson, and R.N. Saunders, *Eur. J. Pharmacol.* **141**, 409 (1987).
42. J. Casals-Stenzel, *Eur. J. Pharmacol.* **135**, 117 (1987).
43. Z. Terashita et al., *Life Sci.* **32**, 1975 (1983).
44. S. Adnot, *Prostaglandins* **32**, 791 (1986).
45. M.S. Wu, T. Biftu, T.W. Doebber, *J. Pharmacol. Exp. Ther.* **239**, 841 (1986).

46. J. Minei, G.T. Shires III, G.T. Shires, *Circ. Shock* **24**, 253 (1988).
47. M. Okamoto *et al., Thromb. Res.* **42**, 661 (1986).
48. A. Etienne *et al., Agents Actions* **17**, 368 (1985)
49. J. Fletcher *et al., Circ. Shock* **27**, 359 (1989).
50. A. Waage, A. Halstensen, and T. Espevik, *Lancet* **1**, 355 (1987).
51. A. Lasfargues and R. Chaby, *Cell Immunol.* **115**, 165 (1986).
52. I.-T. Henter, O. Sodor, and U. Anderson, *Eur. J. Immunol.* **18**, 983 (1988).
53. B. Beutler, I.W. Milsark, and A. Cerami, *Science* **229**, 869 (1985).
54. K.J. Tracey *et al., Nature (London)* **330**, 662 (1987).
55. J.C. Mathison, E. Wolfson, and R.J. Ulevitch, *J. Clin. Invest.* **81**, 1925 (1988).
56. K.J. Tracey, S.F. Lowry, and A. Cerami, *J. Infect. Dis.* **157**, 413 (1988).
57. C.R. Turner *et al., Circ. Shock* **28**, 369 (1989).
58. K.J. Tracey *et al., Science* **234**, 470 (1986).
59. K.E. Stephens *et al., Am. Rev. Respir. Dis.* **137**, 136 (1988).
60. J.R. Weinberg, D.J.M. Wright, and A. Guz, *Clin. Sci.* **75**, 251 (1988).
61. K.J. Tracey *et al, Surg. Gynecol. Obstet.* **164**, 415 (1987).
62. G. Feuerstein *et al., Circ. Shock,* in press.
63. F. Bauss, W. Droge, and D.N. Mannel, *Infect. Immunol.* **55**, 1622 (1987).
64. D.G. Remick *et al., Lab. Invest.* **56**, 583 (1987).
65. C. Nathanson *et al., J. Exp. Med.* **169**, 823 (1989).
66. F. Bussolino *et al., J. Clin. Invest.* **77**, 2027 (1986).
67. F. Bussolino, G. Camussi, and C. Baglioni, *J. Biol. Chem.* **263**, 11856 (1988).
68. H. Hayashi *et al.,* in *Platelet-Activating Factor and Diseases,* K. Saito and D.J. Hanahan, Eds. (International Medical Publishers, Tokyo, 1989), p. 51.
69. B. Bonavida, J.M. Mencia-Huerta, and P. Braquet, *J. Lip. Med.*, in press.
70. B. Bonavida, J.M. Mencia-Huerta, and P. Braquet, *J. Allergy Appl. Immunol.*, in press.
71. M. Rola-Pleszczynski *et al., Prostaglandins* **35**, 802 (1988).
72. X.-M. Sun and W. Hsueh, *J. Clin. Invest.* **81**, 1328 (1988).
73. M. Paubert -Braquet *et al., J. Lip. Med.,* in press.
74. M.G. Sirois *et al., J. Lip. Mediat,* in press.
75. D. Howat *et al., J. Lip. Mediat,* in press.
76. A. Floch *et al., J. Lip. Mediat,* in press.
77. H. Heuer and K.H. Weber, *Prostaglandins* **35**, 814 (1988).
78. A.K. Myers, J.W. Robey, and R.M. Price, *Br. J. Pharmacol.,* in press.
79. G.F. Evans and S.H. Zuckerman *Agents Actions* **26**, 329 (1989).
80. A. Wagga, *Clin. Immunol. Immunopathol.* **45**, 348 (1987).
81. G.J. Bagby *et al., Circ. Shock* **28**, 385 (1989).
82. B. Beutler, I.W. Milsark, and A. Cerami, *J. Immunol.* **135**, 3972 (1985).
83. D.G. Hesse *et al., Surg. Obstet. Gynecol.* **166**, 147 (1988).
84. H.R. Michie *et al., N. Engl. J. Med.* **318**, 1481 (1988).

14

PAF in Renal Physiology and Pathophysiology

correspondence
Gérard E. Plante
Richard L. Hébert
Departments of Medicine and Pharmacology
University of Sherbrooke
Sherbrooke, Québec
Canada

Pierre Braquet
Institut Henri Beaufour
Paris, France

Pierre Sirois
Departments of Medicine and Pharmacology
University of Sherbrooke
Sherbrooke, Québec
Canada

Platelet-activating factor (PAF) is a potent glycero-phospholipid mediator produced in glomeruli and the renal medulla. PAF reduces renal blood flow, glomerular filtration and urinary electrolyte excretion by direct effects on the kidney (renal artery perfusion studies). Potential mediators of these PAF-induced renal effects include prostaglandins and leukotrienes. Significant interaction between PAF and natriuretic hormones (atrial peptide and ouabain-like material) have been described not only in the kidney, but also in specific extrarenal sites, such as the endothelial barrier of pulmonary and intestinal microcirculation and the plasma membrane of vital organs (heart, pancreas, kidneys, and skeletal muscle). Because of the latter effects, PAF can be seen as a potential modulator of internal distribution of body fluid volumes (vascular, interstitial and cellular). A number of recent studies suggest that PAF plays a critical role in a variety of diseases which involve primarily or secondarily specific microcirculation networks. Shock, arterial hypertension, interstitial and cellular oedema, renal ischemic and immune injury, hyperfiltration and progressive renal insufficiency, and transplant rejection represent as many examples in which PAF is presumed to play a critical role. In a number of experimental models of such diseases, PAF antagonists were shown to prevent and/or correct the pathophysiological events involved.

Introduction

Renal production of PAF. The specific CoA-dependent acetyltransferase responsible for the production of PAF from its precursor lysoPAF, as well as the acetyl-hydrolase which inactivates PAF are both present in glomeruli and medullary interstitial cells of the kidney.[1] Schlondorff *et al.*[2] identified the mesangial cell as the main source of PAF-like material in the glomerulus.

Cellular Actions of PAF. Specific membrane PAF receptors have been identified in several vital organs,[3] and the role of intracellular mediators, in particular inositide triphosphate, has also been identified. In isolated endothelial cell preparations, Camussi *et al.*[4] have examined the ultrastructural basis for PAF-induced increment in vascular permeability, and documented the critical role of the cytoskeleton. It is likely that the mesangial cells of the glomerulus operate in a similar fashion.[5]

Effects of PAF on Renal Function

Renal hemodynamics. Renal plasma flow and glomerular filtration are both reduced when small doses of PAF are injected or infused in the dog renal artery. In most circumstances, glomerular filtration decreases more than renal plasma flow, leading to significant reduction in filtration fraction. Since these changes are only obtained in the experimental kidney, it is believed that PAF exerts direct effects on renal hemodynamics.[6] In addition, these changes are entirely reversible within less than sixty minutes (Figure 1). Similar findings were reported by other groups.[7,8]

Renal transport of electrolytes. Urine flow and sodium excretion decrease markedly following intrarenal PAF bolus injection or infusion. As shown in Figure 2, a rebound natriuresis develops at the end of PAF bolus injection, with urine sodium reaching 170 percent of control values at the end of recovery.[6] Fractional excretion of sodium slightly decreases from the control value of PAF, suggesting that PAF might in fact augment the net tubular reabsorption of this cation. In

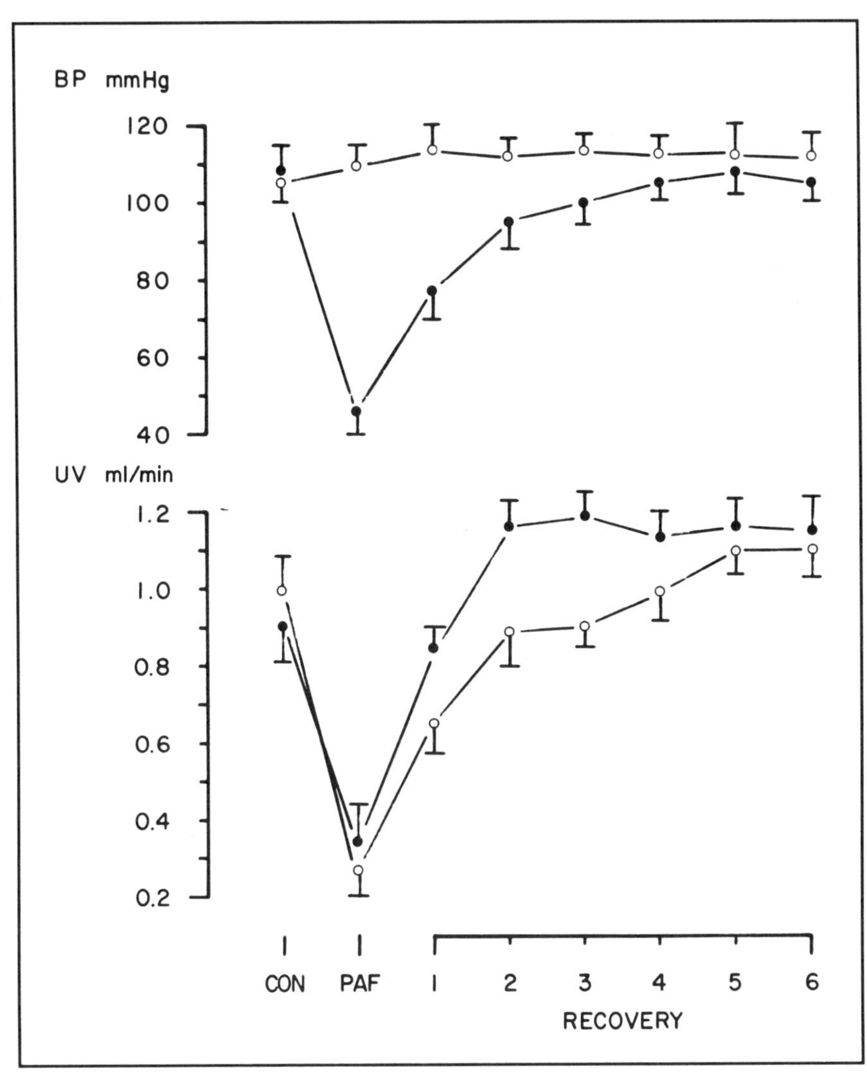

Figure 1. Effect of PAF on systemic blood pressure (BP) and urine flow in control (○) and indomethacin-treated dogs (●). Time periods during recovery are of 10 minutes each.

contrast, during recovery, fractional excretion of sodium increases twofold. As illustrated in Figure 3, changes in blood pressure, urinary sodium, and glomerular filtration induced by PAF infusion are blocked by BN 52021, a PAF receptor antagonist and a valid dose-response curve is obtained.

Figure 2. Effect of PAF on urine sodium (UNaV) and glomerular filtration (GFR). Symbols as in Figure 1.

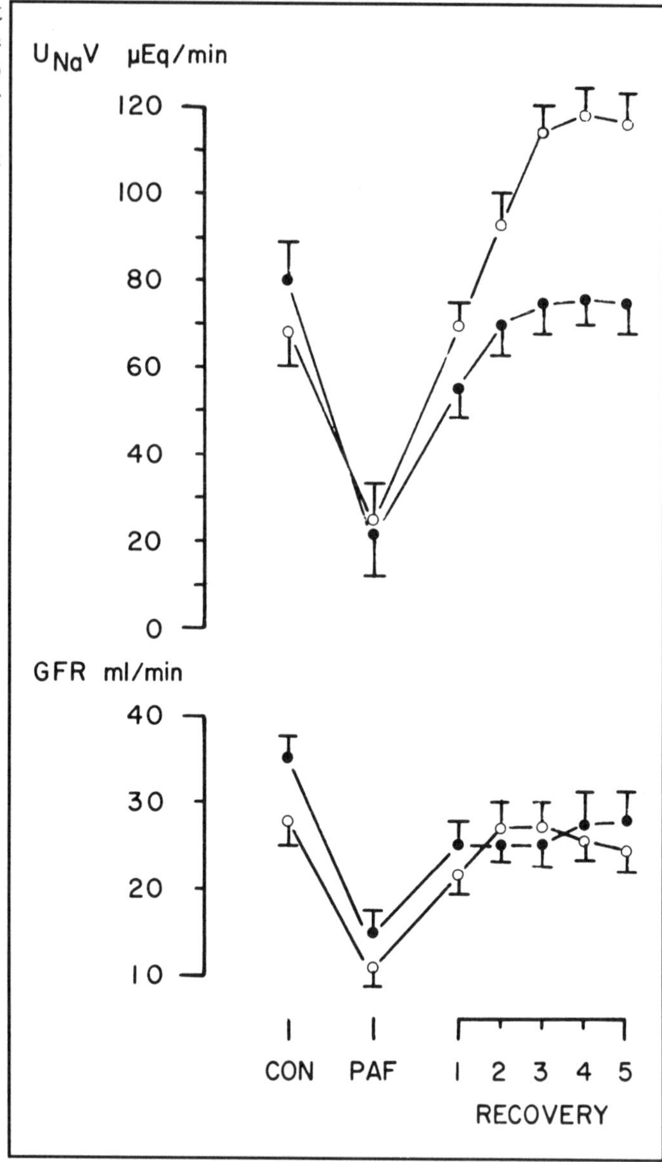

Potential mechanisms. The results obtained in these and other *in vivo* experiments[7,8] indicate that PAF causes a direct effect on the renal vasculature. The variation in filtration fraction, which develops during these changes in renal hemodynamics, probably alters peritu-

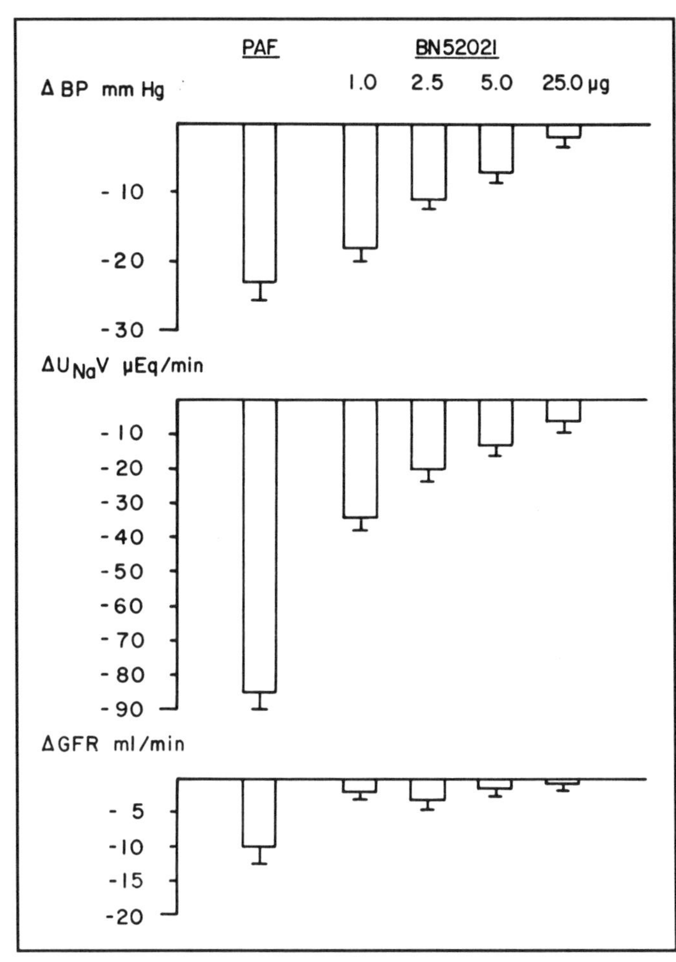

Figure 3. Influence of increasing doses of BN 52021 on the PAF response (blood pressure, urine sodium and glomerular filtration. PAF is administered with each dose of BN 52021, and all changes are compared to baseline values.

bular physical forces, in particular oncotic pressure, which in turn is associated with less passive influx of fluid and electrolytes following PAF injection. However, during the injection of this potent glycerophospholipid, peritubular physical forces and fractional sodium excretion change in opposite directions. It is likely, therefore, that additional mechanisms intervene.

The potential role of other renal autacoids, such as angiotensin II, prostaglandins (including thromboxanes), and leukotrienes has been examined. Saralasin, an angiotensin II receptor antagonist, prevents the PAF-induced reduction of renal plasma flow and glom-

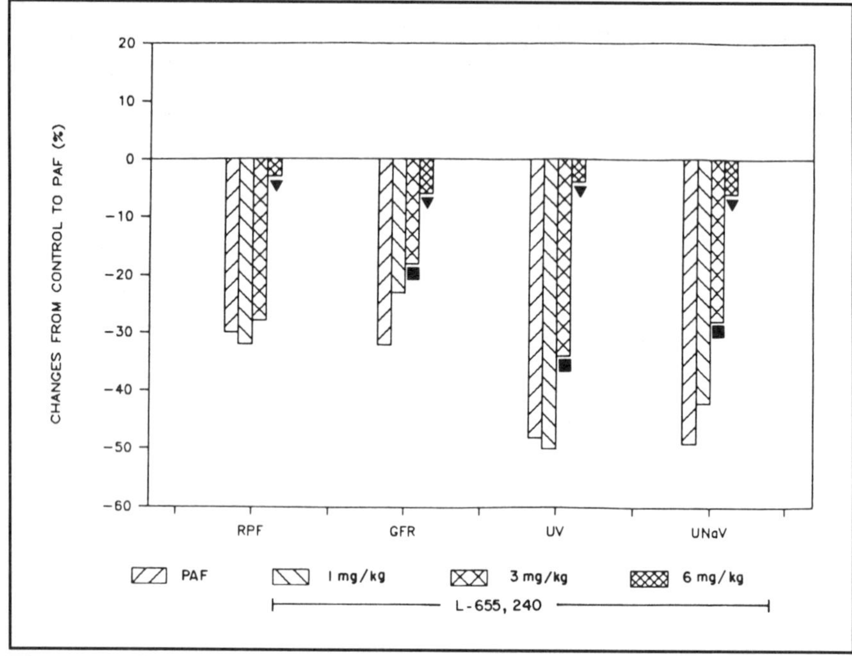

Figure 4. Influence of increasing doses of L-655,240 on PAF-induced changes in renal plasma flow (RPF), glomerular filtration (GFR), urine flow (UV), and urine sodium (UNaV). Procedures and calculations as in Figure 3. ■ and ▼ represent *P* values <0.05 and <0.01, respectively, compared to control values.

erular filtration, indicating, therefore, the participation of this vasoconstrictor peptide in the renal hemodynamic changes that follows PAF injection.[9] Treatment with indomethacin does not prevent the PAF-induced changes in renal hemodynamics. However, the recovery of glomerular filtration following PAF remains incomplete during indomethacin, suggesting the participation of prostaglandins in the adjustement of filtration rate after PAF injection.[6] Similarly, indomethacin prevents the rebound natriuresis that occurs following PAF, again suggesting the role of prostaglandins in this tubular adaptation illustrated in Figure 2. Finally, the potential role of thrombonaxes, and perhaps some leukotrienes as mediators of the PAF-induced renal responses is suggested by the use of L-655,240, a thromboxane/prostaglandin endoperoxide antagonist.[10] Illustrated in Figure 4, this compound was found to inhibit all renal physiological disturbances related to PAF injection in our experimental dog model.

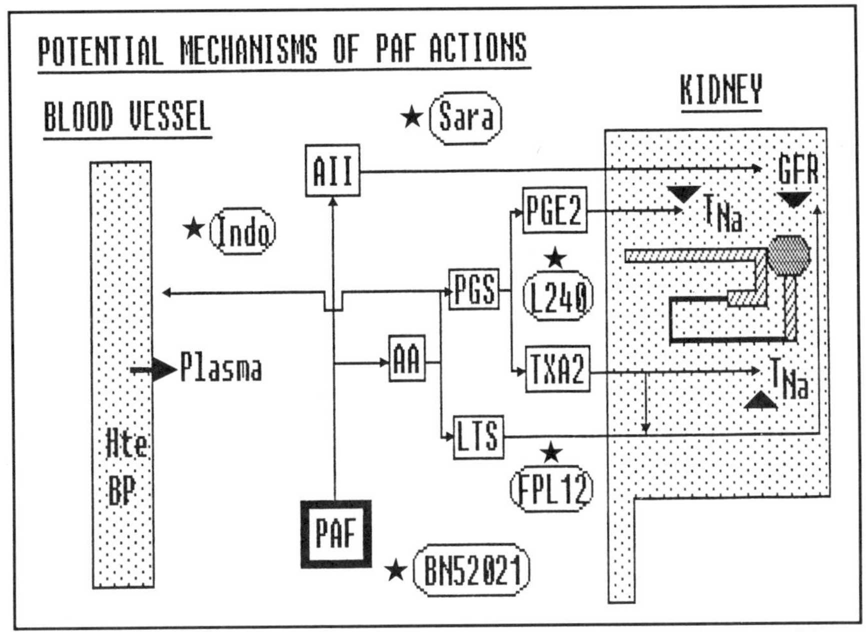

Figure 5. Potential mechanisms involved in the peripheral and renal effects of PAF. Putative mediators are represented in ■, antagonists are indicated by ★.

It is likely that PAF interacts with other endogenous substances known to influence the vasomotor tone and/or sodium transport by the tubular epithelium. Atrial natriuretic peptide and ouabain-like material, both recognized as potent natriuretic hormones, have been shown to interact with PAF and some PAF antagonists in experimental models which examine the extrarenal actions of PAF.[11,12] The eventual role of these compounds in the renal response to PAF has not been directly studied, however. The integration of mechanisms potentially responsible for the PAF-induced renal physiological disturbances described above is presented in Figure 5.

Effects of PAF on Body Fluid Volumes

PAF and vascular permeability. PAF is one of the most potent endogenous compounds capable of shifting plasma from the vascular to the interstitial compartment.[13] When administered intravenously,

PAF may increase hematocrit by more than 50 percent. This effect of PAF on peripheral microcirculation endothelial permeability is responsible for most of the reduction of blood pressure which occurs during PAF injection.[14] Recent experiments from our laboratory, indicate for the first time, that the effect of PAF on endothelial permeability is not uniform. In fact, when injected intravenously in awake normal rats, PAF results in heterogeneous extravasation of Evans Blue, a marker of protein leakage outside of the vascular bed.[15] Pulmonary, gastrointestinal, and pancreatic microcirculation were shown to be mostly affected under these conditions.

PAF during acute saline loading. The most commonly used model to examine the mechanisms of body fluid regulation is the acute response of an experimental subject to acute isotonic saline loading, also termed acute extracellular volume expansion. The term "extracellular" is in fact inexact. During this manoeuvre, the kidney responds in a classical manner: renal plasma flow and glomerular filtration both rise and, more importantly, urinary sodium excretion augments dramatically. The increased natriuresis is secondary to both elevation of filtered sodium and direct tubular inhibition of sodium reabsorption. Changes in glomerulo-tubular physical forces,[16] as well as increased secretion of natriuretic hormones, contribute to the renal adaptive response to expansion.[18]

The extrarenal response to expansion has not been as well described. The renal excretory capacity of an acute saline load is not prompt enough to prevent major changes of the vascular volume. We documented that interstitial and cellular volumes may contribute 50 and 20 percent, respectively, in the "storage" of an acute isotonic saline load.[17] The mechanisms responsible for this extrarenal adaptation are not entirely understood, but indirect evidence suggest that PAF may be one important mediator, shifting part of the saline load to the interstitial compartment, and perhaps, to the cellular volume. It is of interest that the atrial natriuretic peptide possesses similar properties, in addition to its renal effects.[18] It is also likely that the ouabain-like material, also released during expansion, exerts some inhibitory effect on fluid movements from the cellular to extracellular compartments.[19] The proposed interactions between natriuretic

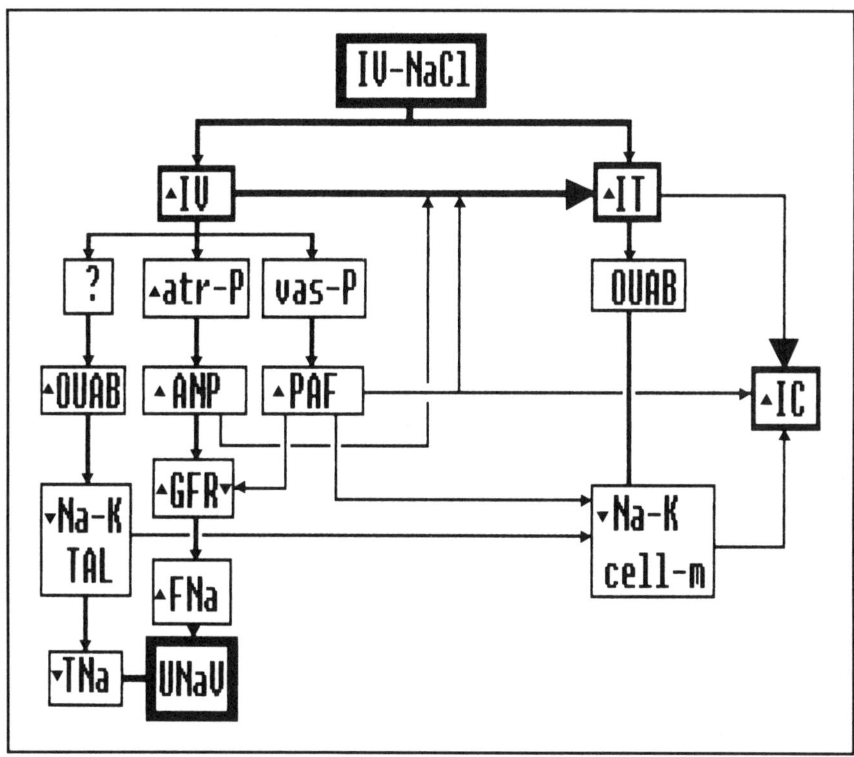

Figure 6. Proposed interactions between atrial natriuretic peptide (ANP), ouabain-like material (OUAB), and PAF during acute isotonic saline loading. Movements of fluid between vascular (IV), interstitial (IT), and cellular (IC) compartments are illustrated.

hormones and PAF during acute saline expansion is illustrated in Figure 6.

PAF as a Mediator of Renal Injury and Related Disorders

PAF in acute renal ischemia. The pathophysiology of acute renal failure has not yet been entirely established, but the role of potent mediators such as PAF has been retained by several investigators. Using a "two-kidney-one-renal-artery occlusion" anesthetized rat model, we recently examined the effect of BN 52021, a selective PAF-receptor antagonist, on the recovery pattern of renal function follow-

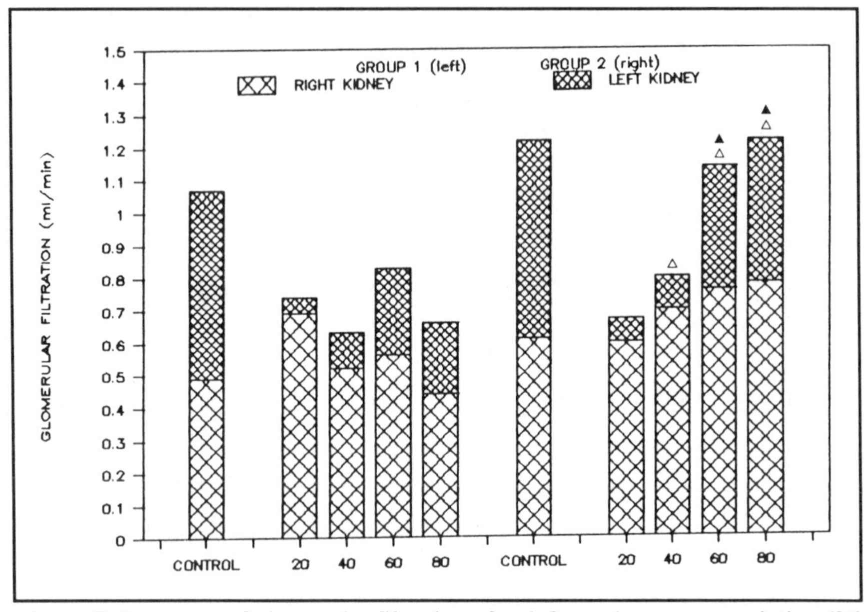

Figure 7. Recovery of glomerular filtration after left renal artery constriction (25 minutes), in control (left side of figure) and BN 52021-treated rats (right side of figure). Values obtained from right (control) and left kidneys (ischemic) are indicated. Recovery is examined over a period of 80 minutes. Δ and ▲ represent P values <0.05 and <0.01, respectively, compared to control values.

ing 30 minutes of complete occlusion of blood flow. As illustrated in Figure 7, eighty minutes after the release of this occlusion, glomerular filtration only recovers to 38 percent of control values in untreated animals, whereas BN 52021-treated rats recover to 69 percent of preischemic values.[20] These results suggest that PAF is involved in the pathophysiology of ischemic renal failure.

PAF in chronic renal failure. The role of hyperfiltration of the remnant "intact" nephron in initiating the development of progressive loss of renal function has been extensively studied over the past decade.[21] Increased glomerular permeability to plasma proteins is an early characteristic of the failing organ under these circumstances. Since PAF increases endothelial permeability to plasma macromolecules, we examined the eventual role of this mediator, using the Bricker model of chronic reduction of the renal mass (4/5 nephrectomy). BN 52021 administered orally during four weeks post nephrectomy did not alter glomerular filtration, when compared to control animals. However, this PAF antagonist prevented the increment in absolute and fractional urinary excretion of sodium, which characterizes untreated animals.[22]

PAF and renal toxic injury. Finally, PAF has also been recently incriminated in the pathophysiology of two types of renal interstitial toxicity of significant clinical importance, cyclosporin and *cis*-platinum nephrotoxicity. Acute and chronic experimental animal models were used to examine the effect of PAF antagonism on the development of glomerular and tubular dysfunctions secondary to cyclosporin and *cis*-platinum. In both models of nephrotoxicity, BN 52021 improved glomerular filtration and histology of injured kidneys, whereas untreated animals developed the classical tubulointerstitial infiltration and late decrease of glomerular filration.[23]

PAF and arterial hypertension. A number of investigators consider that essential and renal-associated hypertension could be related to abnormal distribution of body fluid volumes.[19] The eventual role of PAF as a potential mediator of such disturbances of fluid volume distribution must be examined for two major reasons. First, PAF may be an important regulator of fluid passage between vascular and interstitial compartments, and perhaps, between interstitial and cellular volumes.[17] Second, PAF exerts a selective effect on endothelial permeability, increasing plasma leakage in organs that may be of importance to those organs known to release vaso-active mediators, such as the lung and the gastrointestinal tract.[15] We examined the influence of BN 52021 on blood pressure and renal parameters of "uremic" rats (Bricker model). When administered orally for four weeks, this PAF antagonist significantly attenuated (by 50±3mmHg) the rise of blood pressure that occurs in untreated animals.[22] In addition, BN 52021 was found to diminish the abnormal and selective increment in renal capillary permeability documented in the spontaneous hypertensive rat (SHR) model.[24] It is likely, therefore, that PAF plays a role in some aspects of the complex pathophysiology of arterial hypertension in these models.

Perspectives and Conclusions

Future research in the field of PAF and renal physiology and pathophysiology should include development of PAF agonists and antagonists that may help to understand the cellular mechanisms of action of this important lipid mediator. The clinical and biological tolerance of PAF antagonists should also be examined in carefully designed clinical studies. The use of PAF antagonists in the prevention and even the cure of acute immunological, toxic, and ischemic renal injury seems very close. In addition, such antagonists could eventually be useful in the preservation of organs for transplantation purposes. The role of PAF in the pathophysiology of progressive renal disease is not yet evident, and further research aimed at understanding the mechanisms of hyperfiltration and tubular adaptation to the reduction of the renal functional mass is urgently needed. Finally, the role of PAF in the development of certain types of arterial hypertension deserves special interest, since a number of indirect evidences suggest that this potent glycerophospholipid is an important endothelial autacoid capable of affecting the resistance and permeability of blood vessels, the target organ of this most frequent disorder.

References

1. E. Pirotsky *et al., Lab. Invest.* **51**, 567 (1984).
2. D. Schlondorff *et al., J. Clin. Invest.* **73**, 1227 (1984).
3. E. Kloprogge and J.W.N. Akkerman, *Biochem. J.* **223**, 901 (1984).
4. G. Camussi *et al., J. Immunol.* **131**, 2397 (1983).
5. A.F. Michael *et al., Kidney Internat.* **17**, 141 (1980).
6. R.L. Hebert *et al., Prostaglandins, Leukotrienes, and Med.* **26**, 189 (1987).
7. H. Scherf *et al., Hypertension* **9**, 82 (1986).
8. P. Bessin *et al., Eur. J. Pharmacol.* **86**, 403 (1983).
9. G.E. Plante and R.L. Hebert, in *Gingkolides: Chemistry, Biology, Pharmacology, and Clinical Perspectives,* P. Braquet, Ed. (Prous Science Publishers, Barcelona, 1988) pp. 575-602.
10. R.L. Hebert, P. Sirois, and G.E. Plante, *Can. J. Physiol. Pharmacol.* **67**, 304 (1989).
11. P. Braquet *et al., Pharmacol. Rev.* **39**, 97 (1987).
12. G.E. Plante *et al., Am. J. Physiol.* **253**, R375 (1987).
13. S.-B. Hwang *et al., Lab. Invest.* **52**, 617 (1985).

14. G.E. Plante *et al., Pharmacol. Res. Comm.* **18** (Suppl.173), 345 (1986).
15. M.G. Sirois *et al., Prostaglandins* **36,** 631 (1988).
16. J.W. Lewy and E.E. Windhager, *Am. J. Physiol.* **214,** 943 (1968).
17. G.E. Plante *et al., Hormonal control of body fluid volumes: evidence supporting the role of the atrial natriuretic peptide (ANP), ouabain-like material (OUA), and the platelet-activating factor (PAF).* (Proc. 12th Scientific Meeting of the International Society of Hypertension, Kyoto, May 22-26, 1988).
18. B.J. Ballermann and B.M. Brenner, *J. Clin. Invest.* **76,** 2041 (1985).
19. H.E. DeWardener and E.M. Clarkson, *Clin. Sci.* **63,** 415 (1982).
20. G.E. Plante, P. Sirois, and P. Braquet, *Prostaglandins, Leukotrienes, and Med.* **34,** 53 (1988).
21. B.M. Brenner, T.W. Meyer, and T.H. Hostetter, *N. Engl. J. Med.* **307,** 652 (1982).
22. Y.A. Lussier *et al., Med. Sciences* **5** (Suppl.3), 58A (1989).
23. E. Pirotzky *et al., Adv. Lipid Res.* **23,** 1989.
24. M. Bissonnette *et al., Early selective microvascular disorders in the SHR.* (Proc. 13th Scientific Meeting of the International Society of Hypertension, Montreal, June 24-29, 1990).

Index